U0178532

Create a better living environment for human beings

为人类创造更美好的人居环境

2

赛瑞景观二十周年作品特辑

新共享景观

赛瑞景观 编著

中国建筑工业出版社

序 1

2000 年至今是中国房地产发展至关重要的二十年，赛瑞景观作为第一批参与我国城市建设的景观设计企业，见证并经历了房地产行业的前进与骤变，这是赛瑞景观出版本书的行业发展背景。

我们正在进入高速发展的时代，在城市化大潮的影响下，快速的城镇化进程、人口急剧扩张、共享经济出现，使得景观设计行业与建筑设计行业一样，挑战与机遇并存。共享经济推进了全球化进程，而现代景观设计将面临更多课题：一系列新的经济模式、社会开放关系，城市重构给环境与人类带来了不同程度的负面影响。有效保护地球资源，促进城市发展，成为城市建设者的共同目标，景观设计行业在此扮演着重要角色。

设计的发展趋势将是多元化、世界性的。赛瑞景观在二十年的景观设计实践中，为适应城市发展需求，提出了"新共享景观"设计理念。打破传统的景观设计思路，为未来景观设计行业提出新的研究方向与设计挑战。全球化思考与在地性实践，城市将更注重生态保护与文脉传承。《新共享景观》书籍中记录了赛瑞景观成立二十年来的三十五个实践案例，其中文化遗产建筑修复项目"STAPLE HOUSE"的设计强调与自然共生融合的主张，以最少的设计介入达到历史与现代、人与自然的平衡，很好地诠释了"新共享景观"的概念。

新的理念需要持续实践与时间检验，《新共享景观》是赛瑞二十年与众多同行共同结下的丰硕果实，亦是景观行业二十年发展的城市印记。愿赛瑞景观在更长远的城市景观设计建设中奉献专长，共创下一个十年。

孟建民

2020 年 6 月

PREFACE 1

The 20 years from 2000 until now constitute a key period in the history of Chinese real estate development; as in the first batch of landscape design enterprises participating in China's urban construction, CSC Landscape has witnessed the progresses and leaps of China's real estate industry, which forms the industrial development background for CSC Landscape to publish this book.

As we are entering an epoch featuring rapid development and are under the influence of the tide of urbanization, the fast progress in urbanization, rapid expansion of population, and emergence of sharing economy result in the coexistence of challenges and opportunities in the landscape design industry, like in the construction design industry. The sharing economy promotes the process of globalization, while the modern landscape design will have to confront more problems: a series new economic modes and open social relationships as well as city reconstruction will impose different degrees of negative impacts on the environment and mankind. To effectively protect the earth resources and promote urban development has become the common goal of urban constructors, in which the landscape design industry plays an important role.

The design develops towards a diverse and cosmopolitan trend. In its landscape design practices for twenty years, to meet the needs of urban development, CSC Landscape has put forwards the design concept of *"New Shared Landscape"*. It breaks the conventional thoughts in landscape design, and comes up with new research directions and design challenges for the landscape design industry. Both globalized thoughts and localized practices promote the city to pay more attention to ecological protection and cultural inheritance. *New Shared Landscape* includes 35 practical cases implemented by CSC Landscape for 20 years since its establishment; among them, the "STAPLE HOUSE", a project for restoration of cultural heritage buildings, emphasizes the coexistence and harmony with the nature, and achievement of balance between mankind and the nature with the minimum design intervention, and gives a good interpretation of the concept of *"New Shared Landscape"*. New concepts should be able to bear the continuous test of practice and time; *New Shared Landscape* represents the plentiful fruits harvested by CSC with many partners and also the urban impression left by the landscape industry after its development for twenty years. I hope CSC Landscape can make its contributions to the urban landscape design and construction in a long run and work together to create the next ten years.

Yuezhong Chen

June, 2020

序 2 为赛瑞《新共享景观》序文

喜闻今岁赛瑞景观已经创立二十周年，欢庆之时不忘承启，并将近年来作品去芜存菁，聚诸一书为《新共享景观》，赠书于我并请作序，余不胜喜，欣然从笔。

本书通过广泛的实操案例，从市政、商业、住宅景观全面地呈现了公司作品。书中没有连篇累牍的文字论述，却用翔实的案例引起了读者共鸣，从而引出了赛瑞"新共享景观"的设计理念：

——与自然共生

——与人共享

——与文化共情

景观设计师的修养一定是由环境不断影响着成长的，世界塑造着我们的认知，然后由我们的认知来打造世界；我们以由表及里、由简及繁的项目需求为己任；我们去推敲、去归纳、去想象和推测环境中正在发生的事情，以及需要作出的改善；它的昨天意味着什么？明天又将产生什么变化？破除傲慢与固执后，内外交互影响的过程可以循环不休，直到难分彼此。其时，自然中见天地，人中见众生，文化中见传承，随心所欲而不逾矩，一切都在共生、共享、共情中和谐发展。

二十年来国内的景观行业从星星之火到蔚然大观，从跟随到超越，从学习到布道，在这个恢宏的时代里，万物向阳而生，肆意生长。而赛瑞景观始终保持了一份坚守与执著，持续不断地探索适合自己的道路。《新共享景观》很好地反映了赛瑞景观人的心路历程，也为其他景观人提供了学习与借鉴。

做景观，我们永远在路上。

陈跃中

2020 年 5 月

PREFACE2 Preface to CSC's New Shared Landscape

I am delighted to know that this year is the twentieth anniversary of CSC Landscape, and on such a festive occasion, to create a link between past and future, CSC Landscape has gathered the essence from its past books and created a book named *New Shared Landscape*. I am honored to receive a gift of this book and the invitation to write the preface to this book. Certainly, I accept this invitation with pleasure.

This book offers a comprehensive overview of CSC's works from the aspects of municipal administration, commerce and residential landscapes through an extensive range of real cases. There is no redundant narrative; instead, it chooses complete and accurate cases to evoke the resonance of readers, and then derives the design philosophy of CSC in "New Shared Landscape":

- Coexistence with the nature
- Share with people
- Empathy with cultures

The landscape designers develop their own virtues and qualifications under the incessant influence of environment; as the world shapes our perceptions, we create the world based on our perceptions. We take as our tasks the project demands from the surface to the center and from the simple to the complex. We weigh, conclude, imagine and infer what is happening in the environment and what needs to be improved. What does its yesterday mean? What changes will occur tomorrow? After removing the arrogance and stubbornness, the inside and outside interaction falls into an endless cycle until it is hard to tell one from the other. Actually, you can tell the universe from the nature, see the images of all living creatures in the life of people, and discover the lineage from culture; all things are at their own will while following certain rules, and develop harmoniously in the way of coexistence, share and empathy.

Over the past two decades, China's landscape industry has developed from the sparks of fire to a grand sight, from following to surpassing, and from learning to preaching; in this magnificent epoch, all things on earth grow widely towards the bright sun. CSC Landscape, always with its original insistence, keeps exploring the ways suitable for itself. The New Shared Landscape gives a proper reflection of the mind journey of CSC Landscape, and provide other landscape involvers with learning and borrowing materials.

We are always on the road of landscape practicing.

Yuezhong Chen

May, 2020

前言

文：廖文瑾
　　赛瑞景观董事长

二十年前，赛瑞景观作为中国最早一批将景观规划设计（Landscape design）理念引入中国的开创性专业设计机构，为中国城市园林的建设和发展做出了历史性的贡献！二十年来，赛瑞景观也以自己不断创新的设计理念及众多的优秀作品，参与并见证了中国改革开放的巨大变化及城市建设的日新月异。

作为一家专业的景观设计机构，在不断创新探索的设计道路上，我们赛瑞人切身经历体验到设计创意以及人才团队是何等的重要和宝贵，这已成为我们赛瑞在过去的二十年及至今后的事业征途中，最为重要的课题。

赛瑞景观设计本质的意义就是为人与环境建立一个舒适美好的"关系"，而在这个"关系"里人民对感受的需求和期盼已经越来越大于对功能性的要求，正是基于对这种"关系"的尊重和理解，赛瑞景观在公司成立之际就将"为人类创造更美好的人居环境"为自己的企业使命。

二十年的不断践行探索，赛瑞景观经历了好几个不同的设计理念阶段，而近期在景观行业率先提出的"新共享景观"创新设计概念，也正是赛瑞景观在新的时代，对景观设计的创新探索，同时也希望赛瑞人可以为社会继续呈现更多优秀且有影响力的作品。

没有人才的企业是没有竞争力的，人才的价值就是我们企业的价值，人才的问题是决定我们赛瑞景观未来是否能走得好、走得远的核心问题。二十年来我们赛瑞人经历了太多的失败教训，同时也获得了许多的成功喜悦，正是因为我们赛瑞人的不懈努力和共同追求，才成就了我们今天共同的事业。

正值赛瑞景观二十年庆典之际，我们从过往完成的上千个并在百余城市落地的项目当中筛选出一些具有代表性的优秀作品汇编成集，为的是让赛瑞景观可以做一个阶段性的总结，以便我们在今后的设计创意道路上更上一层楼，同时也请社会各界和行业同仁多点评指正，共同合作发展。

廖文瑾

2020 年 2 月 13 日于加拿大

FOREWORD

Writer: Wenjin Liao
Chairman of CSC

Two decades ago, CSC Landscape, as the first batch of pioneering professional design organizations in China to introduce the concept of Landscape design into China, made historic contributions to the construction and development of Chinese urban landscape! Over the past twenty years, CSC Landscape has also participated in reform & opening up, and witnessed great changes and rapid development of urban construction with its own innovative design concepts and numerous excellent works.

As a professional landscape design organization, we have experienced the great significance and value of design creativity and talented team on the road of continuous innovation and exploration in design, which has been becoming the most important topic in our career.

The essential meaning of CSC Landscape design is to establish a comfortable and beautiful "relationship" between human beings and environment, however, people's demand and expectations for feelings have become greater than the requirements of functionality. Based on respect and understanding of this "relationship", CSC Landscape always adheres to "creating better living environment for human beings" as our mission, since the establishment of the company.

During twenty years of continuous practice and exploration, CSC Landscape goes through several stages with different design concepts. The innovative design concept of *New Shared Landscape*, recently first proposed in the landscape industry, is also CSC Landscape's innovative exploration of landscape design in the new era. Colleagues also hope that CSC people can continue to present more excellent and influential works for the society.

An enterprise without talents is not competitive, and the value of talents is the value of the enterprise. The talent is the core factor that determines whether CSC Landscape can go well in the future. Over the past twenty years, we CSC people have experienced so many failures and lessons, and our colleagues have also gained many joys of success. It is because of our unremitting efforts and common pursuit that we have achieved our common cause today.

Upon the twentieth anniversary celebration of CSC Landscape, we select representative and excellent works from thousands of projects which were completed in more than a hundred cities in the past and compile them into a collection, in order to make a periodic summary and do better on the road of design creativity in the future. At the same time, we also invite colleagues from all the walks of life to review, correct and cooperate for development.

Wenjin Liao

February 13, 2020 in Canada

目录

16

ATALOGUE

新基章景观

引言 INTRODUCTION

计之道，
于从社会互动、人文活力、
态可持续发展的综合角度，
予场地新的活力，
于此，
们提出了 ——“新共享景观”设计理念。

—— 丁炯

ne idea of design is to endow the site with new vitality from the
mprehensive perspective of social interaction,
umanistic vitality and ecological sustainable development.
ased on that,
e put forward a new concept of "*New Shared Landscape*" design.

—— Jiong Ding

生态景观 - 与自然共生 - *SYMBIOSIS*

互动景观 - 与人共享 - *SHARED*

主题景观 - 与文化共情 - *EMPATHY*

新共享景观

文：丁炯
赛瑞景观董事总经理

一、源起

1. 环境问题

在全球化的今天，随着科学技术的不断革新和发展，城市化进程大幅加速，使得城市与自然愈加疏远。应地，因人口与工业的聚集所引发的人类生活、社会经济与自然、生态环境间的矛盾也愈加突出。同城市化进程也助长了城市的同质化倾向，碎片式与多元文化结合的冲击也使得城市景观渐渐失去自有独的文化属性。更加严重的是，城市化还带来了一系列亟待解决的城市病：交通拥挤、环境污染、秩序乱……我们眼中鳞次栉比的城市建筑，行色匆匆的街头人流，高墙矗立的封闭社区，让人与社会、人与之间竖起了一道无形的墙。

2. 思想意识

随着我国经济社会的发展，共享理念成为现代经济的一种共同的理念。共享单车的盛行，共享汽车、共充电宝以及共享雨伞等系列服务产品应运而生，"共享经济"作为一种新经济形式，被各行业学习、借和运用，共享经济的影响越来越深。实践证明，它有利于解决城市问题，有利于提高公众生活满意度。

二、理念

设计发展之道在于认识存在的问题、直面存在的问题，利用不断发展的科学技术与思想认识，回应新的境与社会诉求，突破传统模式，不断注入新的活力。基于此，我们提出了新共享景观设计理念。

新共享景观包含三个方面内容：与自然共生（生态景观）、与人共享（互动景观）、与文化共情（主题观）。

三、释义

1. 与自然共生（生态景观）

如果把景观设计理解为是一个对人类使用户外空间及土地的分析、提出解决问题的方法以及监理这一解方法的实施过程，那么景观设计从本质上是对土地和户外空间的生态设计，是一种基于自然系统自我机更新能力的再生设计。"与自然共生"，我们认为包含两个层次的内容。其一，关于自然本身。我们要去保护与修复它，降低新建项目对环境的不利影响，并提升改造项目场地恢复的能力。遵循地域性原自然性原则、边缘性原则、多样性原则去进行生态性设计。我司在加拿大温哥华的 STAPLES HOUSE 文遗产修复项目就是其中的代表性作品。STAPLES HOUSE 位于西海岸，其设计以"侘寂"为灵魂概念，重对自然生境及场地的保护与恢复，强调新建部分"轻轻地触碰海岸"，最终达成了新、旧两部分与大然的共生共融。其二，人与自然的关系。我们反对唯生态设计，我们主张人与自然互动。鼓励人在与自的互动中，去学习与反馈，去不断平衡与自然的关系。

2. 与人共享（互动景观）

人是社会的动物，而当下社会普遍存在人口高流动性的特点，这削弱了人与人之间的情感维系，形成了对独立的个体。而互联网的出现代替了实体社交空间的地位，在虚拟网络中满足了人与人之间联系性求，因此出现"聚而离"的社会状态。然而人，始终是现实的产物，我们终究要回到现实中来。我们希人与人能发生美好交流与互动，我们需要人与人有情感连接。"与人共享"的设计理念便应运而生，它含三方面的内容。其一，空间共享。城市是一个综合体，目前的规划设计更多地强调其边界，而忽视其合。我们不仅需要关注红线内的设计，也需要着眼于提升边界空间的互动性。比如我们可以将社区红线行退让，把原本失落的社区内边角空间变成城市公共活力空间，形成街角口袋公园，提升社区活力，展现对城市的关怀。或者我们推广共享社区花园，让高层住户也有机会一圆花园梦。其二，功能共享。我们不再进行单一的极致功能布局，而是让部分功能更加复合，强调人与人的交流与共融，更融洽地拉近彼此的距离。比如我们针对现代社区用户需求，创新性地提出三大社区复合公共空间——共享中心、乐享中心和怡享中心，分别关注社交、儿童老人游憩、多功能运动健身，营造全新的未来社区空间格局。其三，信息共享。现代科技发展迅速，信息共享的时代早已到来，信息共享以互动景观作为载体，使人与人互动、线上线下互动。

互动景观作为"与人共享"理念的一个集中体现，景观与人的关系从传统的单向映射到今天的互动，体现的是景观设计理念的革新和科学技术的发展。从单纯的对景观的视觉感受和使用，向依据使用者的感受、反馈进行自我调整和适应。通过运用多种手段激发人的感官体验，让人参与到景观中去，这是互动景观的主要发展方向和目标。我司将互动景观作为新景观探索的一个重要方向，成立互动景观工作室，把研究成果与实际项目相结合。目前已经创作并落地两个项目，分别是湘潭长房·万楼公馆展示区和深圳大鹏海之韵。两者均得到当地政府、业主方的高度认可，到访群众以非常高的热情参与其间，改变了公众对传统景观的认识。

3. 与文化共情（主题景观）

在快速的城市化进程中，现代化的城市设计理念促使城市的发展呈现出同质化倾向，而承载城市历史文脉的景观文化地位岌岌可危。如何保护和整合现有城市景观资源，建构激发大众情感体验和审美需求的公共空间，同时保存代表城市历史的文化标签，是我们不断研究的课题。"与文化共情"设计理念，主张在乡村与都市、历史与当代、地域性与全球化三个维度，寻找与我们的在地文化锚固的共情点，以主题景观的形式表达出来。使我们的景观更加具有地域性和归属感，也更加能反映当代人的历史文化情怀。比如我们的长沙国际会议中心景观设计便是在寻求项目本身与湖湘文化共情点的基础上，再结合建筑"水云万象、山水洲城"的理念，提出"月满星城、山水湖湘"的总体概念。在此概念的指导下，各分区再以细分的共情点予以落地。特别是屋顶花园，我们选取湖湘文化的代表——岳麓书院作为共情点，以现代的手法表达其精神，延续其文脉。

新共享景观设计理念是我们因应环境的变化、时代的发展提出的对策，我们希望其所包含的三者之间是相互支持、相互融合的，它们共同促进人与自然、人与人、人与社会的交流与互动，共融与共享。

2020 年 2 月 25 日 于深圳

New Shared Landscape

Writer: Jiong Ding
Managing Director of CSC

I. Origin

1. Environment Issue

In today's globalization, with the continuous innovation and development of science and technology, the process of urbanization has accelerated significantly, making cities and nature more and more alienated. Correspondingly, the contradiction between human life, social economy, nature, and ecological environment caused by the aggregation of population and industry has become increasingly prominent. At the same time, the urbanization process has also contributed to the homogeneity of the city. Due to the impact of fragmentation and multiculturalism, the city landscape has also gradually lost its unique cultural attributes. What more serious is that urbanization has brought a series of urgent urban diseases: congested traffic, environmental pollution, and chaotic order.From our perspective, row upon row of urban buildings, rushing streets, and closed communities with high walls, make an invisible wall between people themselves and society.

2. Ideology

With the development of China's economy and society, the concept of sharing has become common among modern economy. The prevalence of shared bicycles, a series of service products such as shared cars, shared portable battery, and shared umbrellas came into being. As a new economic form, the "shared economy" has been learned and used by various industries. The impact of the sharing economy is deepening. Practice has proved that it is conducive to solving urban problems and improving public life satisfaction.

II. Idea

The way of design development lies in recognizing the existing problems, facing them, using the continuously developing science and technology and ideological knowledge, responding to new environmental and social demands, breaking through the traditional model, and constantly injecting new vitality. Based on this, we propose a design concept of new share landscape.

The new shared landscape contains three aspects: symbiosis with nature (ecological landscape), sharing with people (interactive landscape), and empathy with culture (themed landscape).

III. Interpretation

1. Symbiosis with Nature (Ecological Landscape)

If the landscape design is an analysis of human use of outdoor space and land, a solution to the problem, and the implementation of the solution which is supervised, then the landscape design is essentially the ecological design of the land and outdoor space, and it is a regenerative design based on the organic self-renewal ability of the natural system. From our point of view, the "Symbiosis with nature" contains two levels of content. Firstly, it is about nature itself. We need to protect and repair it, reduce the negative impact of new projects on the environment, and improve the ability to restore the project site. Follow the regional, natural, marginal, and diversity principles for ecological design. One of our representative works is the STAPLES HOUSE cultural heritage restoration project in Vancouver, Canada. STAPLES HOUSE is located on the west coast. Its design is based on the concept of "silence", focusing on the protection and restoration of natural habitats and sites, and emphasizing that the newly-built part has a gentle relationship with the coast. In the end, the symbiosis between the new and old and nature was achieved. Secondly, the relationship between human and nature. We are opposed to eco-only design. We advocate human interaction with nature. We encourage people to learn in their interaction with nature, and to constantly balance the relationship with nature.

2. Sharing with People (Interactive Landscape)

Human beings are the animals of society, and the current society is characterized by high population mobility. It has weakened the relationship between people and formed relatively independent individuals. The emergence of the Internet has replaced the status of the physical social space. In the virtual network, the need for connectivity between people has been met, so a social state of "gathering away" has emerged. However, people can not live without reality, and we must return to it after all. We need good communication and interaction between people, and we need emotional connection. The design concept of "sharing with people" came into being, and it contains three aspects. The first one is space sharing. The city is a complex, and the current planning and design emphasizes its borders and ignores its integration. Not only do we need to focus on the design within the red line, we also need to focus on improving the interactivity of the boundary space. For example, we can concede the red line of the community and turn the corner space of the originally lost community into an urban public vitality space, thus to form a street corner pocket park to enhance the vitality of the community and show concern for the city. Or we promote the sharing of community gardens, so that high-rise residents' garden dreams can come true. The second one is function sharing. We no longer carry out a single ultimate function layout, but make some functions more complex, emphasizing communication between people, and make them feel closer to each other. For instance, we have proposed three major composite public spaces with many shared centers and centers for enjoyment, which focus on the needs of social activities, recreation for children and the elderly, and multifunctional sports. Create a new future community space pattern. The third one is information sharing. Modern technology is developing rapidly, and the era of information sharing has already arrived. Information sharing takes interactive landscape as a

carrier to enable people to interact with each other, online and offline. Interactive landscape is a concentrated expression of the concept of "sharing with people". The relationship between landscape and people is mapped from the traditional one-way to today's interaction, which reflects the innovation of landscape design concepts and the development of science and technology. It changes from the simple visual perception and use of the landscape to self-adjustment and adaptation based on the user's feelings and feedback. The main development direction and goal of interactive landscape is letting people participate in the landscape by using various methods to stimulate people's sensory experience. Our company regards interactive landscape as an important direction for new landscape exploration, and we established an interactive landscape studio to combine research results with actual projects. At present, two projects have been created and implemented, they are the Chanfine · Wanlou Mansion Exhibition Area in Xiangtan city and Shenzhen Dapeng Haizhiyun. Both were highly recognized by the local government and the owner. The common people participated with great enthusiasm and they are the game changing about the public's understanding of traditional landscapes.

3. Empathy with Culture (Theme Landscape)

In the process of rapid urbanization, modern urban design concepts promote the homogeneity of urban development, and the landscape cultural status that bears the city's history is at stake. The subject of our continuous research is how to protect and integrate the existing urban landscape resources, build public spaces that stimulate the public's emotional experience and aesthetic needs, and at the same time preserve the cultural label that represents the city's history. The design concept of "Empathy with culture" advocates to find the empathy points anchored with our local culture in the three dimensions of village and city history and contemporary regionality and globalization, then express it in the thematic landscape. That can make our landscape more regional and belonged, and also reflect the historical and cultural feelings of contemporary people. For example, the landscape design of our Changsha International Conference Center is based on the project's own sympathy with Huxiang culture, combined with the concept of "Water, clouds, mountains around the city", and proposes the overall concept of "Starry city with mountains and rivers around". Under the guidance of this concept, subdivisions will be implemented. Especially for the roof garden, we choose Yuelu Academy, which represents the Huxiang culture, as the empathy point. We express its spirit in a modern way and keep its culture alive.

The concept of new shared landscape design is our response to the changing environment and time. We hope that the three elements are mutually supportive and integrated with each other since they jointly promote the interaction between human and nature, people and society and people themselves.

February 25, 2020 in Shenzhen

新景观的生态性表达

文：何国平
赛瑞景观设计总监

历史的发展过程，解决发展问题的同时伴随着发展产生的新问题。我们在创造了丰富的物质文明的同也带来了一系列生态环境问题。因此解决好人与自然的和谐共处，有效地利用自然资源，营造良好的可续的生态环境，将是未来景观设计的主题。共享景观在与自然的对话中，力求满足人们对环境的需求，成文化的表达及传承，通过景观设计寻求一条人与自然之间的共生之路。

与自然共生之"生态性"

探寻人与自然之间的关系，首先要探讨景观设计当中生态性的问题，而解决生态性问题的首要前提是尊自然。因此，我们要寻找到人和自然之间的平衡。与此同时，在设计过程中也要通过景观生态性的方法决现有的生态环境问题，在每一个设计的场所中植入生态理念和具体的实施办法。共享景观理念是在尊自然的基础上追求景观生态性的表达，为人和场所之间创造出一种可持续的共生状态。

与自然共生之"美学性"

共享景观理论也置于景观生态化的美学表达，表现自然之美是景观设计中的重要内容，设计是创造美的程，好的景观设计作品应该满足生态需要和使用功能，同时也是一件美的艺术品。因此我们在追求景生态性的基础上，也要在视觉上让作品呈现美的状态。美和生态是共存的，两者的结合才是我们所追求从探讨"与自然共生"的概念出发，共享景观理念通过研究不同生态元素间的相互作用，将灵感转化合成为多变的空间、色彩和肌理。在设计过程中出于审美需求我们更倾向于偏向自然的表达，比如材料身的自然呈现更具本真性的张力，同时我们也会使用能够衬托自然之美的材料，比如透明和半透明的具时代性性质的材质。这种材料的半透明性可以很好地呈现自然光线的变化，另外它可以在人工和自然之间生一种互动状态，透过材料表现自然万物的趣味性。

与自然共生之"互动性"

与此同时，我们认为共享景观可以通过连接人和自然以提升人们的场地参与感，丰富人们生活的同时，强场地活力。同时鼓励人们外出活动，以多种多样的生活体验促进人们与场所之间的互动。这也是实现地可持续性和弹性的重要手段。 真正关注并实现了环境的塑造与提升，不仅与人们需要的功能空间完结合，同时尊重本土独特的自然条件。

与自然共生之"文化性"

共享景观的理念不仅探索自然与人的关系，而且有意去探索一种文化属性的自然表达。受中国传统山水化的影响，对我们来说人是自然的一部分，并且乐于在日常生活中和自然互动。当我们设计景观时，希通过共享景观理念强调这种具有文化属性的自然关系，通过对空间的营造真正能做到寄情于"山水"之我们不是独立于自然之外，我们本身就为营造环境而存在，做好人与自然之间的平衡就是我们所追求这种共生状态：与自然共生，与文化共情，与人共享。这既是共享理念的主旨，也是我们赛瑞人的不懈求！

2020 年 2 月 12

Ecological Expression of New Landscape

Writer: Guoping He
Design Director of CSC

Looking at the historical development process, we can found that the new problem always arises when the development problems were solved. While creating a rich material civilization, we have also brought a series of ecological and environmental issues. Therefore, it is necessary to maintain the harmonious coexistence between man and nature, effectively use natural resources, and create a good sustainable ecological environment. That will be the theme of future landscape design. Shared landscape can meet people's environmental needs, complete the culture expression and inheritance, and seek a symbiotic path between man and nature through landscape design.

"Ecology" that Coexists with Nature

To explore the relationship between human and nature, we must first explore the ecological issues in landscape design, and the primary prerequisite for solving these issues is the respect for nature. Therefore, we need to find a balance between man and nature. At the same time, in the design process, the existing ecological environment problems should be solved through ecological methods, and ecological concepts and specific methods should be implanted in each design site. The shared landscape concept is to pursue the expression of landscape ecology on the basis of respect for nature, and to create a sustainable symbiotic state for people and sites.

"Aesthetics" Symbiosis with Nature

The shared landscape theory is also placed in the aesthetic expression of landscape ecology. The expression of natural beauty is an important content in landscape design. Design is the process of creating greatness. Good landscape design works should meet both the ecological needs and functions, and meanwhile be a great art work. Therefore, based on the pursuit of landscape ecology, we must also visually make the works beautiful. Beauty and ecology coexist, and the combination of the two is what we seek. From the concept of "symbiosis with nature", shared landscapes transforms and combines inspiration into changing spaces, colors, and textures by studying the interaction between different ecological elements. In the design process, due to aesthetic needs, we tend to prefer natural expressions. For example, the natural material itself has more authentic tension. At the same time, we also use materials that can set off the beauty of nature, such as transparent and translucent materials that can reflect the time. The translucency of this material can well represent the change of natural light. In addition, it can create an interactive state between the artificial and nature, and express the interest of nature.

"Interactivity" Symbiosis with Nature

At the same time, we believe that shared landscapes can enhance people's sense of site participation by connecting them with nature, also it can enhance the vitality of the site while

enriching people's lives. And people are encouraged to go out to promote interaction between people and sites with a variety of life experiences. That is also an important means to achieve site sustainability and resilience. It paid real attention to achieve the shaping and improvement of the environment. It is not only perfectly combined with the functional space that people need, but also respected the unique natural conditions of the locality.

"Cultural" Symbiosis with Nature

The concept of shared landscapes not only explores the relationship between nature and human, but also deliberately explores the natural expression of a cultural attribute. Influenced by Chinese traditional landscape culture, human beings are a part of nature for the Chinese, and they are willing to interact with nature in daily life. When we design landscapes, we hope to emphasize this kind of natural relationship with cultural attributes through the concept of shared landscapes. Through the construction of space, we can truly be in love with "landscape".

We, humans, are not independent of nature, we exist to create the environment, and the balance between man and nature is what we seek. This symbiotic state: symbiosis with nature, sympathy with culture, and sharing with people. That is not only the main theme of sharing ideas, but also the unremitting pursuit of our people!

February 12, 2020

体验式互动景观策略探索

文：徐伟瀚
赛瑞景观项目副总监

科技发展日新月异，人对景观的需求不再止于纯粹的观赏，对景观的情感体验提出了更高的要求，"与人共享"的景观设计理念由此应运而生。互动景观突破传统景观的界限，从功能、艺术、体验等多方面出发为人类生活创造了更多的可能性。互动景观强化乃至重构了景观与个人、空间与行为之间的关系。新时代人们追求的生活不仅仅是舒适和健康，更多的是追求多元化的生活体验，"与人共享"作为一种新的景观设计理念，更好地指导人对景观的一些实际需求及解决人与景的关系问题。

在国内现有的互动式景观建设案例中，大部分设计理念及建设施工等都处于起步和摸索阶段。设计着重美感，容易忽视人们实际使用的便捷性和趣味性，使得国内的互动式景观难以形成一套行之有效的设计方法。与此同时，目前行业内存在着景观在设计上千篇一律的问题，导致设计对使用者的吸引力逐渐降低。因此在未来的设计中要在观念上强调场地与使用者的双向映射，实现真正的人与景观的互动。

国外的互动式景观注重人的使用体验，强调通过科技的手段、产品化的设计实现人与景观的互动。在景观设计中，人的需求被充分考虑，例如针对弱势群体的需求进行专项研究及关怀化设计。但是，过分依赖互动装置，也会缺少对整体景观设计的把控，需要对体验式景观提出一定的设计策略，避免未来呈现产品化的趋势。

基于上述现状及未来需求，我梳理出以下三种互动式景观设计策略。

"诱发性"景观策略

日常行为活动或游戏的激发需要具备一定的诱发因素。想要诱导人的行为体验，需要首先创设一个能够诱发人感知的体验主题。在设计互动体验式主题景观时，可以根据设计的主题预测在此景观空间中是否会激发与主题相关的自发活动，来延长人在空间中的逗留时间，即可以大大加强人与景观的互动体验性。

"技术性"景观策略

多媒体技术与景观的融合能改变景观的形态、色彩，改变人对景观的感受，增强景观与人的互动，创造出人与人互动的情境。"技术性"景观主要分成两种：一是多媒体为主导介质的景观，另一种是多媒体辅助介质的景观。

"智慧性"景观策略

利用现代智能新技术，使景观空间作品具有拟人化的智慧，能根据人的需要，对输入的信息做出合适的反应，与人产生积极的互动。互动体验式景观可以利用"智慧"理念，把现代智能技术植入景观作品中，实现现场信息、个人现场体验感受与互联网展示空间信息同步，大大增加景观对人的吸引力。

事实上，实现与人共享，既需要有良好的策略和景观大局观，也需要结合科技手段，实现景观的产品化、功能化，这样才能真正体现出互动式景观的本意。因此与人共享的互动交流性引发了人们强烈的参与感，这不仅仅是现代景观设计中很重要的一个设计手段，更是未来景观设计发展的新趋势。

在这个休闲娱乐经济的时代，与人共享的互动体验式景观是一种关注人心理需求的景观理念，通过我们的共同努力，最终达到与人共享、与自然共生的和谐状态。

2020 年 2 月 20

Exploration of Experiential Interactive Landscape Strategy

Writer: Weihan Xu
Deputy Project Director of CSC

With the rapid development of science and technology, people's demand for landscapes is no longer limited to pure viewing, and higher demands are made on the emotional experience of landscapes. The concept of "sharing with people" has showed up as the times require. The interactive landscape breaks through the boundaries of traditional landscapes, and creates more possibilities for human life from the aspects of function, art, and experience. Interactive landscape strengthens and even reconstructs the relationship between landscape and individual, space and behavior. In the new age, people's pursuit of life is not only about comfort and health, but more about the pursuit of a diversified life experience. "Sharing with people" as a new concept of landscape design can better guide people's actual needs for landscape and solve the problem about the relationship between people and landscape.

Among the existing interactive landscape construction cases in China, most of the design concepts and construction are at the initial stage. The design focuses on aesthetics, and it is easy to ignore the convenience and fun of people's actual use, making it difficult for domestic interactive landscape to form a set of effective design methods. At the same time, there is a problem of uniformity in landscape design in the industry, which has gradually reduced the attractiveness of design to users. Therefore, in the future design, it is necessary to conceptually emphasize the mutual mapping between the site and the user to achieve the real interaction between people and the landscape.

Interactive landscapes abroad focus on the human experience, emphasizing the use of technology and product design to achieve interaction between people and the landscape. In landscape design, human needs are fully considered, such as special research and caring design for the needs of vulnerable groups. However, excessive reliance on interactive devices also lacks control over the whole landscape design. It is necessary to propose a certain design strategy for the experiential landscape to avoid the trend of productization in the future. Based on the status quo and future needs mentioned above, I sort out the following three interactive landscape design strategies.

"Evoked" Landscape Strategy

The stimulus of daily behavioral activities or games requires certain triggering factors. In order to induce people's behavioral experience, we must first create an experiential theme that can stimulate people's perception. When designing an interactive experiential theme landscape, you can predict whether or not spontaneous activities related to the theme will be triggered according to the landscape, thus to extend the person's stay in the space, which can greatly enhance the interactive experience between people and landscape.

"Technical" Landscape Strategy

The integration of multimedia technology and landscape can change the form and color of

the landscape, change people's perception of the landscape, enhance the interaction between the landscape and people, and create more situations where people interact with each other. There are two main types of "technical" landscapes: one is the landscape with multimedia as the dominant medium, and the other is the landscape with multimedia as the auxiliary medium.

"Smart" Landscape Strategy

Using new modern intelligent technology, landscape works shall have anthropomorphic wisdom, and can respond appropriately to input information according to people's needs, and have positive interactions with people. Interactive experiential landscape can embed modern intelligent technology into landscape works, synchronize on-site information, personal on-site experience and information on the Internet display space and greatly increase people's attractiveness of the landscape by using the "wisdom" concept.

Actually, to achieve the concept of sharing with people, it is necessary to have a good strategy and overall landscape view, as well as the combination of scientific and technological means to achieve the productization and functionalization of the landscape. Only by that the original intention of interactive landscape can be truly reflected. Therefore, the interaction and sharing with people has triggered a strong sense of participation. This is not only a very important design method in modern landscape design, but also a new trend in the future development of landscape design.

In this era of leisure and entertainment economy, the interactive experiential landscape is a landscape concept that pays attention to the psychological needs of people. Through our joint efforts, we can achieve a harmonious state of sharing with people and coexisting with nature.

February 20, 2020

景观设计中的文化引入

文：李烨
赛瑞景观设计副总监

文化是一种社会现象，又是一种历史现象。它体现在地理历史、风土人情、传统习俗、生活方式、文学术、哲学思想等各个方面，是不同地域民族之间普遍认同的可持续发展的一种意识形态。

景观设计与文化之间相互作用，相互影响。景观是文化的物化和载体之一，文化是景观进化和演变的内文化特质被铭刻于景观设计之中，从文化角度去塑造景观。而在文化的传承与弘扬过程中，景观又成了们的心灵寄托和情感归宿。

随着经济的发展，人们已不再追求温饱与生存，更注重品质与体验的提升，对文化及精神层面有着更多追求。景观设计作为一种媒介，能否准确传达场地文脉，使人感知，让使用者与场地产生联想与火花，文化产生共情，都需要我们从场地文脉、历史文化、使用者体验、时间维度等多方面进行思考和设计。

首先，在城市公共景观空间中，设计者需要了解场地和使用者，尊重场地的历史文脉和特点，了解场地身的逻辑和秩序，并且探讨使用者会对哪些元素产生共情反应，或者对哪些元素产生认知感、归属感和密感。除了这些，还需要人们主动参与其中，发生互动，真切感受景观设计带来的文化内涵。让人与城文化、自然建立关系，产生共情。湖南省博物馆、长沙国际会议中心两个项目就将"鼎盛洞庭"的历史脉，以及"山水洲城"的城市特征，融入到了景观中，将湖湘文化的底蕴、三湘四水的城市文脉生动地示了出来，从而提高了城市空间的凝聚力和认同感，顺应了城市发展的时代需求。

其次，我们在挖掘场地文化，以及产生共情的潜质因素的基础上，运用了一定的设计技巧，使其文化价被展示放大，从而让人从内心深处与场地产生联系，唤醒人们内心的温暖记忆。设计者深入了解现状场的构成元素、秩序和情境。引用原场地的元素进行设计创新，充分延续场地的地域文脉，带给场地新的密度和归属感。只有尊重场地文化特色，才会被人们所认同。长房时代公馆项目就是以湘潭书院为文化景，用现代手法展现了湘学始源的博厚悠远，延续了场地书香文脉和烟雨塘的精神印记，使人们与场地忆和城市文化产生共鸣。

再次，设计者在传承历史文化精髓时，也应注重景观与文化的时间维度。因为随着时代的变迁，人的心审美习惯会发生改变，一味地照搬旧的文脉和设计语言是不能满足当代人需求的，这就要求景观给文脉入新的血液元素，注重使用者的感受，符合现代人的使用习惯、精神追求，引导人与文化产生共情。永安建发玺院项目便是不拘泥于传统形式和材料，通过现代设计手法，将古典造园精髓及古人所追求"结庐在人境，而无车马喧"的生活理想化入了景观之中，使传统文化与现代功能达到了水乳交融的境这就是对传统文化理解后的设计创新，是对美好人居环境向往的真情流露。

最后，景观设计师在传承与弘扬文化的同时，从文化的发展中去理解景观，理解文化与人的关系，与场的关系。在设计中，把使用者对于场地的体验、感受和需求与场地本身的潜质性和精神放置在同一条线激发人与环境、人与文化之间产生共鸣与火花，推动文化与现代景观设计的融合与发展，在实践中探寻化与景观设计之间的平衡点、创新点。

2020 年 2 月 24

Cultural Introduction in Landscape Design

Writer: Kevin Lee
Deputy Design Director of CSC

Culture is a social phenomenon and historical phenomenon. It is reflected in geographical history, local customs, traditional customs, lifestyles, literature and art, philosophical thoughts and other aspects. It is an ideology of sustainable development generally recognized among different regions and nations.

Landscape design and culture interact and influence each other. Landscape is one of the materialization and carriers of culture, which is the connotation of landscape evolution. Cultural characteristics are engraved in landscape design, and the landscape is shaped from a cultural perspective. In the process of cultural inheritance and promotion, landscape has become the spiritual sustenance and emotional destination of people.

With the development of the economy, people are no longer pursuing subsistence, they are paying more attention to the improvement of life quality and experience, and they are pursuing more in the aspect of culture and spirit. As a medium, whether landscape design can accurately convey the context of the site, make people feel it, let users associate with the site and empathize with culture, all require us to think and design from many aspects such as the context of the site, history and culture, user experience, time dimension and so on.

First of all, in the urban public landscape space, designers need to understand the site and users, respect the historical context and characteristics of the site, understand the logic and order of the site itself, and explore which elements the user will have empathic reactions to, or which elements create a sense of cognition, belonging, and intimacy. In addition to these, people also need to actively participate in it, interact, and truly feel the cultural connotation brought by the landscape design. The designer needs to connect people with city, culture and nature, so that they can empathize with them. The two projects of Hunan Provincial Museum and Changsha International Conference Center have integrated the historical context of the "Prosperous Dongting" and the urban characteristics of the "Landscape City" into the landscape, integrating the heritage of Huxiang culture and the urban culture of Sunshine Four Rivers. The pulse is vividly displayed, thereby improving the cohesion and identity of the urban space, and conforming to the needs of urban development.

Secondly, on the basis of excavating the culture of the site and the potential factors that generate empathy, we have used certain design techniques to display and amplify its cultural value, so that people can connect with the site from the heart and the warm memories of theirs will be awaken. The designer has a deep understanding of the elements, order, and context of the site, introducing elements of the original site for design innovation, fully extending the local context of the site, bringing new intimacy and a sense of belonging to the site. It is the design which respects the cultural characteristic of the site that can be respected by people. The Changfang Times Mansion Project is based on the cultural background of

Xiangtan Academy, using modern methods to show the richness and longevity of the origins of Hunan academics, and continues the spiritual imprint of the site's sacred context and Yanyutang, so that people can resonate with the site's memory and urban culture.

Thirdly, when inheriting the essence of historical culture, designers should also pay attention to the time dimension of landscape and culture. Because with the changes of the times, people's psychology and aesthetic habits will change as well. Blindly copying the old context and design language cannot meet the needs of contemporary people. This requires the landscape to inject new elements into the context, pay attention to the user's feelings, meet the modern people's habits, spiritual pursuits, and guide people to empathize with culture.

Yongan Jianfaxiyuan Project is not limited to traditional forms and materials. Through modern design methods, the essence of classical gardening and the pursuit of ancient people's "to settle in the realm of people, but without car and horses" are idealized into the life. In the landscape, the traditional culture and modern functions have reached a state of perfect harmony. This is the design innovation after understanding the traditional culture, and it is the true feelings for the beautiful living environment.

Finally,while inheriting and promoting culture, landscape architects understand the landscape from the development of culture, the relationship between culture and people, and the relationship with the site. In the design, the user's experience, feelings and needs for the site are placed on the same line with the potential and spirit of the site itself, creating resonance between people and environment, culture while integrating culture and modern landscapes. The balance and innovation between culture and landscape design are explored in practice.

February 24, 2020

生 态 景 观 - *S Y M B I O S I S*

生态环境的恶化，使得生态成为当代景观设计领域的重要议题之一。景观设计师掌握着将环境政策和社会审美转译成建成环境的途径，因而在积极面对来自人居环境、栖息地、水环境等生态环境问题时，透过"与自然共生"的景观设计理念，不断平衡人与自然之间的关系，激活场地的活力，满足生态可持续发展。

With the deterioration of ecological environment, ecology has become one of the major issues in the field of contemporary landscape design. Landscape designers master the approach of transforming environmental policies and social aesthetics into the built environment. They can constantly balance the relationship between human beings and nature, activate the vitality of the site, and meet the ecological sustainable development via the landscape design concept of "symbiosis with nature", when they actively face eco-environmental problems resulted from human settlements, habitats, water environment and so on.

Staples House 的修复与建设，设计方在设计上更注重的是保留原有西海岸风情的建筑风格以及建筑主材，并在此基础上增加新的功能设计。

Staples House 位于加拿大西海岸温哥华，是由世界建筑大师亚瑟 · 埃里克（Arthur · Erickson）设计并修建于 20 世纪 60 年代的一座代表作品，现已被列为 West Vancouver 文化遗产建筑，该作品曾作为 2015 现代建筑展六大展示作品之一展出。由于 Staples House 的历史性和建筑代表性，业主及温哥华政府一同合作，进行修复建设，并提供详细方案，最终打造成温哥华具有重要历史意义的建筑典范。该项目是由赛瑞景观董事长廖文瑾先生带领其项目团队主持的历史文化遗产修复项目，并对此进行了深入的文化挖掘及研究设计。

Regarding the restoration and construction of the Staples House, the designer pays more attention to the design of retaining the original West Coast style of the building and the main building materials, on the basis of which new functional design is added.

Located on the West Coast of Canada, the Staples House is a masterpiece designed by Arthur · Erickson, a master architect in the world, and was constructed in 1960s. It has been listed as the cultural heritage of West Vancouver. The work has been exhibited in 2015 Chinese Contemporary Architecture Exhibition as one of the six major exhibition works. Based on its historicity and architectural representativeness, the owner teamed up with Vancouver municipal government to restore and construct this architecture and provided detailed schemes to build it into an example of important buildings in Vancouver. Mr. Liao Wenjin, the Chairman of CSC Landscape, led his team to carry out this historical and cultural heritage restoration, and carried out in-depth culture mining and research design.

STAPLES HOUSE 文化遗产修复项目
The Cultural Heritage Restoration Project of STAPLES HOUSE

设计时间：2015 年
项目地址：加拿大西海岸温哥华
项目类型：文化遗产修复项目
设计单位：加拿大赛瑞 (CSC) 设计顾问公司
　　　　　/ Nick 建筑师事务所

Design time: 2015
Location: The west coast of Canada
Project type: The cultural heritage restoration project
Design unit: CN Shine Consulting Co., Ltd.
　　　　　 / Nick Milkovich Architects Inc.

地区背景

在加拿大，对历史建筑修复及建设需要政府、建筑师、开发商和民众，包括监管和操作员等共同参与商讨。每个人都可以发表自己的观点，大到建筑理念小到每一项材料的选择，这样的"会议"从开始设计到完工会持续进行下去。政府、建筑师、开发商、民众，在城市历史建筑保护与修建中，谁也不能缺位，城市历史建筑保护与修建切勿"丢掉了城市自己的痕迹"。历史遗产建筑的保护和更新已成为中外都十分重视的课题。北美的历史没有中国丰富悠久，但是他们似乎更重视对历史的保护——不论是历史文化，还是历史建筑。加拿大在历史建筑保护工作中已建立严谨的法律法规，城市建设流程必须确保建筑可持续性，对于 Staples House 的保护修建项目的重视和实践，也可以为我国历史遗产建筑的保护和更新提供一定的经验。

定位策略

这个位于悬崖上的建筑，以一种特殊的形式致敬自然：它轻轻地触碰海岸，传递着对自然的赞美。每一座历史建筑都反映着当地的历史文化，承载着生活在这里的人和环境的故事。城市建设急促发展的步伐，逐步带走了历史遗留的记忆。历史建筑是一个民族、一个地区乃至整个国家的文化历史的物质反应，是文明进步的见证者，对其加以妥善保护和完成传承就是延续历史。赛瑞一如即往地承担起对社会历史的责任，关注历史建筑修复与建设的动态，并进行持续研究，助力推进历史建筑保护进程。

Background

In Canada, the government, architects, developers and the public, including regulators and operators, are required to participate in the repair and construction of historical buildings. Everyone can express their own views, ranging from the architectural concept to the choice of each material. Such "meetings" will continue from the beginning of design to the completion. No one can be absent from the protection and construction. The protection and construction of the urban historical buildings must not "lose the traces of the city itself". The protection and renewal of historical heritage buildings have become a very important topic both at home and abroad. North America's history is not as rich and long as China's, but they seem to pay more attention to the protection of history, whether it's history and culture, or historical buildings. Canada has established strict laws and regulations in the protection of historical buildings. The urban construction process must ensure the sustainability of buildings. The emphasis and practice on the protection and construction of the Staple House can also provide some experience for the protection and renewal of historical buildings in China.

Positioning Strategy

This cliff building honors nature in a special way: it touches the coast gently, conveying a compliment to nature. Each historical buildings reflects the local history and culture, bearing the story of the people and environment living here. The urban construction which develops rapidly has gradually taken away the historical memory of the old city. Historical architecture is the material response of culture and history of a nation, a region and the whole country. It is the witness of civilization progress. To properly protect its inheritance is to continue history. CSC will take on the social responsibility of the enterprise as soon as possible, continue to pay attention to the dynamic of the restoration and construction of historical buildings, and carry out continuous research to help facilitate the protection process of historical buildings.

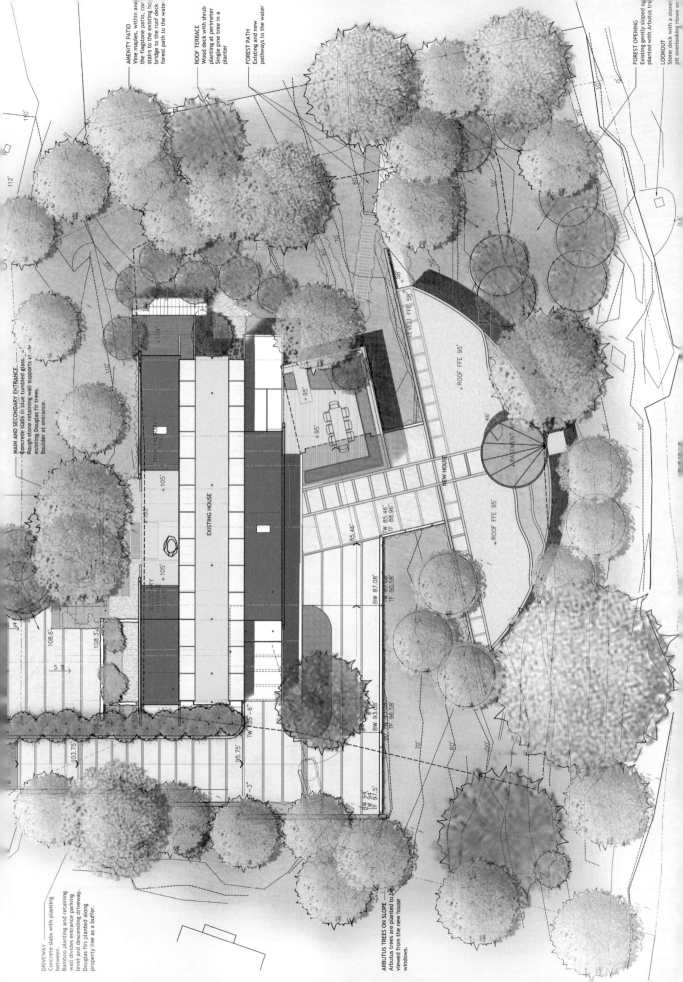

DRIVEWAY
Concrete slabs with planting between.
Bamboo planting and retaining wall divides entrance parking level and descending driveway.
Douglas firs planted along property line as a buffer.

ARBUTUS TREES ON SLOPE
Arbutus trees are planted to be viewed from the new house windows.

MAIN AND SECONDARY ENTRANCE
Concrete slabs in blue tumbled glass.
Rough stone retaining wall supports el...
existing Douglas fir trees.
Boulder at entrance.

AMENITY PATIO
Vine maples, within an...
the flagstone patio, co...
stairs to the existing ho...
bridge to the roof deck
forest path to the water

ROOF TERRACE
Wood deck with shrub planting at perimeter
Single pine tree in a planter

FOREST PATH
Existing and new pathways to the water

FOREST OPENING
Existing gently sloped o...
planted with Arbutus tre...

LOOKOUT
Stone deck with a stone...
pit overlooking Howe so...

EXISTING HOUSE

NEW HOUSE

+ 105'
+ 105'
+ 105'
+ 95'
+ 95'
+ 95'
+ ROOF FFE 95'
+ ROOF FFE 95'
+ BASEMENT...
LEVEL FFE 58'
+ 58'
+ 46'
+ 104'

MAIN ENTRANCE...
SECONDARY ENTRANCE FFE 105'

TW 105'-6"
BW 93.68'
TF 96.58'
BW 87.08'
TF 90.58'
BW 94'
TW 94'
TF 97.5'
TW 85.46'
TF 88.96'
85.46'
95.75'
103.75
108.6'
108.3'
112'
110'
116'
5 %
70'
60'
50'
30'
20'
10'

设计理念

"侘 · 寂"是贯穿整个设计的灵魂概念。侘 · 寂发音为"Wabi Sabi",描绘的是残缺之美。协调旧建筑与新建筑的融合,是 Staples House 项目对"Wabi Sabi"概念的诠释。Staples House 的修复与建设,赛瑞在设计上更注重保留建筑原有西海岸风情的建筑风格同时,还融入了新的功能设计。

建筑设计

在 Staples House 的修复设计中,除了对 1 号房子（旧的文化遗产建筑）进行优化设计,设计方在 1 号房子下方,悬崖上,向下延伸出三层高建筑,即为 2 号房子（扩建项目）,它的设计创意为"灯塔"。"灯塔"的设计遵循了 HRA 条例,这样的改造策略能做到将用户观景体验达到极致的同时,也将对建筑未来创造更多的经济效益,更有助于遗产建筑的维护及可持续发展。

Design Concept

"Wabi Sabi" is the soul concept throughout the whole design. "Wabi Sabi", pronounced Wabi Sabi, depicts the beauty of imperfection. It harmonizes old buildings with new ones, which interpret the concept of Wabi Sabi by the Staples House project. For the restoration and construction of the Staples House, CSC focuses on the design of retaining the original West Coast style of the building, on the basis of which new functional design is added.

Architectural Design

Therefore, in the restoration design of Staples House, in addition to the optimization design of House 1 (the historical buildings of cultural heritage), the designer extended a three-story building, namely House 2 (expansion project), which we call "Lighthouse", under House 1 and on the cliff. Lighthouse" is designed in accordance with HRA regulations. Such transformation strategy can achieve the ultimate user viewing experience, create more economic benefits for the future of the building, and contribute to the maintenance and sustainable development of heritage buildings.

▲ 1 号和 2 号房子模型效果图
No. 1&No.2 House Model Renderi

◀ 1 号和 2 号房子模型图
No. 1&No.2 House Model

▶ 1 号房子实景图
No. 1 House Real Map

Staples House 的 1 号房子，坐落在一个单一且独特的坐标维度上：在西海岸沿岸上。Staples House 西海岸风格梁柱木骨架的结构及无框大玻璃设计诠释着典型北美西海岸的建筑风格，这也是"轻轻地触碰海岸"的理念最直观的表达。

2 号房子，"灯塔"坐落在悬崖的岩石上，在 1 号房子下方。这个的方位设置，既不影响到 1 号房子的功能及观景视野，对邻居的观景效果也不会产生任何的影响。

"灯塔"设计

"灯塔" 是由水平线性结构组成，这座建筑通过四个圆柱垂直屹立在悬崖表面。然而，他并不是 Staples House 的 1 号房子的衍生品，钢铁和玻璃结构的现代风格也在此体现了设计的升华。除了钢结构构成，设计提取了 Staples House 标志性的设计元素——屋顶天窗。建筑外墙采用大面积白色玻璃包裹，最大限度做到建筑与环境的有机结合。

"灯塔"犹如一个窥视着世界的镜头，通透的弧形镜面为建筑遮风挡雨，让人在室内尽情地享受阳光的沐浴。它延续 Staples House 的设计理念——"轻轻地触碰海岸"。通过最小的改变，让场地变得更漂亮并赋以新的意义。在整体的规划设计中，两座建筑在设计与功能上达到互补，在设计风格上统一而又有变化。

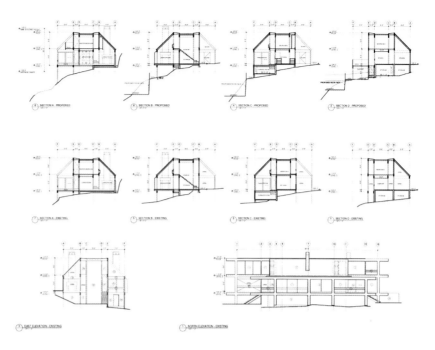

▲ 1 号房子剖面图
Sectional View of the No.1 House

▶ 1 号房子实景图
No.1 House Real Map

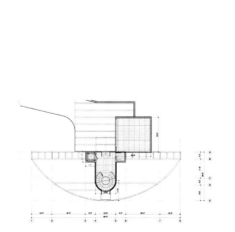

▲ 2 号房子平面图
Floor Plan of the No. 2 House

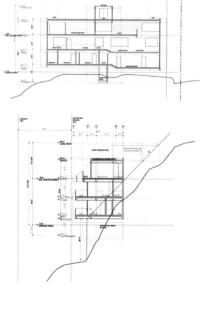

▲ 2 号房子剖面图
Sectional View of the No. 2 House

House 1 of Staples House is located on a single and unique coordinate dimension: along the West Coast. The Staples House's structure of wooden framework of beam-column and the design of frameless large glass best illustrate the typical architectural style of North America's West Coast, which is the most intuitive expression of the concept of "gently touching the coast". House 2, "Lighthouse" is located on the rock of cliff, under House 1. The setting of this orientation will not affect the function and view of House 1, nor the view of neighbors.

"Lighthouse" Design

"Lighthouse" is composed of horizontal linear structure. The building stands vertically on the cliff surface through four columns. However, it is not a derivative of the Staples House 1, where the modern style of steel and glass structures sublimates the design. In addition to the steel structure, the design extracts the iconic design element of the tower house - roof skylight. The outer wall is wrapped with white glass, which can achieve the organic combination of building and environment to the greatest extent.

"Lighthouse" is like a lens peering at the world. The transparent arc-shaped mirror is used to shield the wind and rain in the room and let people enjoy the invasion of sunlight. It continues Staples House's design philosophy: "touch the coast gently", making the site more beautiful and giving new meaning through the smallest change. In the overall planning and design, the two buildings complement each other in design and function, and the design style is both unified and varied.

景观设计

景观设计的主旨依然是"恢复"。设计上仅采用本地植物物种，只有这样的原有物种才能在如此困难的悬崖环境中自然生长。在整个修复设计及建设的过程中也尽量避免对此文化遗产建筑造成破坏。

Landscape Design

The theme of landscape design is still "restoration". Only local plants are used in the design, in this way, such primitive species can grow naturally in such a difficult environment. In the whole restoration design and construction, we avoid causing damage to this cultural heritage.

号房子实景图
5 1 House Real Map

生态景观强调人类生态系统内部与外部环境之间的和谐，系统结构和功能的耦合，过去、现在和未来发展的关联，以及天、地、人之间的融洽性。其中人与自然共生、回归自然、贴近自然、自然融于城市是生态景观的核心内容。

The ecological landscape emphasizes the harmony between the internal and external environment of human ecosystem, the coupling of system structure and function, the connection of past, present and future development, and the harmony between heaven, earth and human. Among them, symbiosis between human and nature, return to nature, close to nature, and integration of nature into the city are the core contents of ecological landscape.

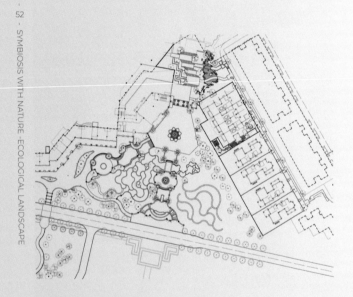

深圳泰富华天峦湖
Shenzhen Topfar Tianluan Lake

设计时间：2013 年
项目委托：泰富华集团
项目地址：广东 深圳
项目规模：135000 ㎡

Design time: 2013
Entrusting party: TOPFAR Group
Project site: Shenzhen, Guangdong
Project scale: 135, 000 ㎡

▲ 七彩池 Colorful Pool

▲ 展示区鸟瞰图
Aerial View of the Exhibition Area

基址位于深圳坪山新区大山坡水库东南侧，基地内地势平缓，与周边地势有一定高差。除了坐享坪山日渐繁华的商业之外，更坐拥一山、三湖、四公园的湖山资源，可谓与山水为邻，私属阔景尽享无余。周边更有红花岭水库、矿山水库环伺，生态环境受到国家法律严格的保护。

The base site is located in the southeast of Dashanpo reservoir in Pingshan New District, Shenzhen. The terrain in the base is gentle, with a certain height difference from the surrounding terrain. In addition to enjoying the increasingly prosperous business of Pingshan, there are also such natural resources as one mountain, three lakes and four parks, gorgeous private broad scenery. There are Honghualing reservoir and Mine reservoir around, and the ecological environment is strictly protected by national laws.

① Water Micro - Climate-Ecosystem
② Blooming-Petal Shaped Waterscape
③ Budding - Flower Bud Planting Pond
④ Sprouting - Vine Landscape Wall
⑤ Mountain Range with Seven Colors - Deducing Natural Form in Space
⑥ Axis Landscape - Creating Natural Elements in the Technique
⑦ The Colorful and Secluded Environment - Creating a Natural Environment Ecologically

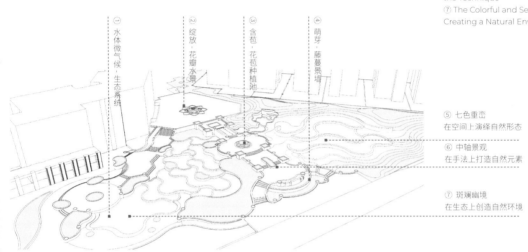

① 水体微气候·生态系统
② 绽放·花瓣水景
③ 含苞·花苞种植池
④ 萌芽·藤蔓景墙
⑤ 七色重峦 在空间上演绎自然形态
⑥ 中轴景观 在手法上打造自然元素
⑦ 斑斓幽境 在生态上创造自然环境

项目以与自然共生为设计主导，以"现代手法诠释自然"的设计手法，模仿天然绝境五彩池与自然跌水，打造出都市少见的天然水系景观体验，向自然致敬的同时，以开放的态度与其和谐共处。
整个空间与周围的环境和谐共融，引导人们暂离喧嚣的城市生活走向集自然、惬意为一体的绿色景观空间，在近自然的项目背景下，给予居者与自然充分交流的空间，于天地宽广中尽情感受自然的意趣。

The project takes the symbiosis with nature as the design guide and "modern interpretation of nature" as the design method. It imitates the natural colorful pool and the natural water falling, creates the rare natural water system landscape experience in the city, pays homage to the nature and harmoniously coexists with it with an open attitude.
The whole space is harmonious with the surrounding environment, guiding people temporarily away from the noisy urban life to a green landscape space integrating nature and comfort. Under the background of near nature projects, the space for residents to fully communicate with nature is provided, and the natural interest is fully felt in the broad world.

中轴景观 Axis Landscape

▲ 中轴景观 Axis Landscape

中轴景观

整个中轴景观以自上而下的自然水系来串联，用跌水与花阶来体现山与水的融合、静和动的搭配，去繁从简，营造人们在山水自然中的生态宜居环境。

通过自然蜿蜒的小径穿行于起伏有致的山地和密林，贯通泳池和溪流，组合成的一曲轻松惬意的园林乐章，令人行走在园中时，处处感受着大自然的拥戴，如在画中，在都市里实现山居生活的生命力，使人在其中生活的状态得以丰盈，赋予景观更丰厚的精神营造。

七彩池

主入口两侧打造自然生态的七彩池与七彩花田，以绝境九寨沟五彩池为设计灵感，在山水相融理念上进行延伸，摒弃传统的几何线条，用有机的自然曲线表达自然万物的气韵流动。一山一水，相得益彰，使主入口景观更具观赏性和感染力。

七彩池池底材料采用蓝色玻璃马赛克，使水流在池中倒映出更加丰富的色彩变化，以人造方式重现大自然的鬼斧神工。

Axis Landscape

The whole central axis landscape is connected by the natural water system from top to bottom. The combination of mountain and water, the combination of static and dynamic, the simplification of complexity, and the creation of an ecological and livable environment for people in the nature of mountains and rivers are reflected by water falling and flower steps. Through the natural winding path through the ups and downs of mountains and dense forests, through swimming pools and streams, the composition of a relaxed and comfortable garden movement makes people feel the support of nature everywhere when walking in the garden. The mountain life is realized in the city, which makes people's living conditions abundant and endows the landscape with richer spiritual construction.

Colorful Pool

Both sides of the main entrance create a natural and ecological colorful pool and colorful flower field, which take Jiuzhaigou Wucai Pool as the design inspiration, extend the concept of landscape integration, abandon the traditional geometric lines, and express the natural charm and flow of all things with organic natural curves. One mountain and one water, reflecting each other, make the main entrance landscape more ornamental and appealing.

The bottom material of the seven color pool adopts blue glass mosaic, which makes the water flow reflect more colorful changes in the pool, and recreates the masterpieces of nature in an artificial way.

▶ 七彩池 Colorful Pool

珠海半岛景观以休闲体验式的时尚、热带风情、都市、伊甸园为设计理念，以轻松休闲的慢半拍节奏，甩开工作与生活中的琐碎，迈入充满生机与活力的都市绿洲，尽情享受幸福生活。运用现代自然园林手法处理"虚与实"的自由多变空间，营造能够触动视觉的设计色彩。景观色彩力求稳重、简洁大气、具有亲和力，整体以浅色为主，以深色混色为辅，亭、廊构筑以深咖色为主，使之更为沉稳亲切。

The Zhuhai Peninsula is designed with the concepts of experiential fashion, tropical amorous feelings, urban and the Garden of Eden. With the leisurely pace of slow half-beat, it is free from the trivial in work and life, and it enters the urban oasis full of vitality and vigor. The landscape makes tourists enjoy a happy life to the fullest. The landscape uses the modern natural garden techniques to deal with the free and changeable space of "virtual and actual", creating the design colours that can touch the visual.
The landscape color strives to be stable, concise, and approachable. The overall color is mainly light color, supplemented by dark color mixing, and the pavilion and corridor are mainly with dark coffee color, making it more calm and friendly.

珠海蓝湾半岛 6-10 期
Herathera Peninsula in Zhuhai #6-10

设计时间：2015 年　　　　　Design time: 2015
项目委托：世荣兆业　　　　　Entrusting party: SHIRONG ZHAOYE
项目地址：广东 珠海　　　　　Project site: Zhuhai, Guangdong
项目规模：115363 ㎡　　　　　Project scale: 115, 363 ㎡

61 · 与自然共生·生态景观

赛瑞景观二十周年作品特辑

6-8 期设计理念

入口空间与周围的环境和谐共融，通过强调现代感的轴线关系对空间进行营造，引导人们暂离喧嚣的城市生活走向集自然、人文于一体的绿色景观空间，聆听水声潺潺，徜徉于枝干与叶片间，感受斑驳的阳光倾洒的惬意，享受自然带来的极致生活体验。

入口桃花和镜面跌水，呈现清波倒影之景，两侧植栽带及拼缝景墙界定边界，群落式种植，塑造风情的丛林空间，区分边界的同时隐约透出其后的浓浓绿意，伴随竹风阵阵，彰显简约和文艺格调。斑驳的竹影在虚实结合的景墙上作画，随着光影的变化，画面丰富而多变。于休闲广场中静坐，看庭前花开花落，望天空云卷云舒。

▶ 小区入口水池
Community Entrance Pool

he entrance space is harmonious with the surrounding environment. By emphasizing the construction of the space by the

:is relationship of the modern sense, people are temporarily guided away from the noisy urban life to the green landscape

ace integrating nature and humanity, listening to the sound of the water, wandering between the branches and leaves,

eling the satisfaction of the mottled sunshine, and enjoying the ultimate life experience brought by nature.

: the entrance, peach blossom and mirror water fall, presenting the reflection of the waves. Planting belts and mosaic walls on

oth sides define the boundary, community planting, shaping the forest style, at the same time, faintly showing the following

ick green, accompanied by the bamboo wind, highlighting the simplicity and artistic style. The mottled bamboo shadow

painted on the scenery wall with the combination of the virtual and the real. With the change of the light and shadow, the

cture is rich and changeable. Sit in the leisure square, watch the flowers blooming and falling in front of the court, and watch

e clouds rolling in the sky.

- 63 · 与自然共生·生态景观

▲ 中心景观
Central Landscape Area

马蹄踏水 - 中心景观区

马蹄形水面围合草坪设计，岛状植栽、乔木草坪、背景植栽形成对景，立面层次丰富。水景崇尚自然，采用大体量的水体，于浅水面戏水、草坪中静坐，动静相宜间，将戏水的乐趣无限延展放大。水面到地面过渡自然，几何草坪呼应方正院落空间，利用大曲线与平缓的大草坪起伏，形成流动的韵律，结合镜面水景提供静思赏景的空间。

棕榈树阵打造浓郁的东南亚风情，错落有序的层次，创造别样的穿行体验，为居住者打造空间、建筑、景观与生活四者充分交融的度假社区。

通过常绿组团，利用植物机理营造院落的层次感，运用公园尺度打造热带丛林体验，设计利用原路穿行于椰林树影的空间中，增加步行体验多层次的趣味感受，营造舒缓的音乐节奏。

Horseshoe Treading on Water (Central Landscape Area)

Horseshoe shaped water surface enclosed lawn design, island planting, arbor lawn, background planting form the opposite landscape and rich facades. The waterscape advocates nature, adopts large water body, and sits in shallow water, lawn, with appropriate movement and stillness, so as to enlarge the fun of swimming. The transition from the water surface to the ground is natural. The geometric lawn responds to the square yard space. The large curve and gentle fluctuation of the lawn are used to form a flowing rhythm. The mirror water view is used to provide a space for contemplation and appreciation.

Palm tree array creates a strong Southeast Asian ethos, scattered and orderly levels, creating a different travel experience, and creating a fully integrated holiday community, architecture, landscape and life for residents.

Through the evergreen group, the use of plant mechanism to create a sense of level of the courtyard, the use of park scale to create a tropical jungle experience, the design directs the original road to travel through the space of coconut tree shadow, to increase the multi-level experience of walking, creating a soothing music rhythm.

9-10 期设计理念

流水蝉声（主入口）

入口空间丰富的细节和生动的自然情趣共同组成了多层次的空间，使园境成为统一纯粹的整体，营造出一种古朴的、淡雅的、神秘的感受，方便人与自然之间的对话。

疏林蜜意（中心景观区）

中央花园以开阔的阳光草坪为主要空间，四周结合疏林草地、台地花园、林荫广场、下沉广场等各具特色的花园空间，有机的路网串联各个空间，动静结合，相得益彰。园区内采用浓密的植物、天然木材、石头和雕塑充满了原始淳朴地域的色彩，一石一木，或含蓄或静谧，曲折、平缓的节点意境融合相间，简练直接而不失细腻委婉，让游园之人悠闲惬意、静谧清闲。

▲ 主入口 Main Entra

0 Phase Design Concept

ater Fall with Cicada Sound (Main
trance)

e rich details and vivid natural
ements of the entrance space form
multi-level space, which makes the
rden a unified and pure entirety,
eating a simple, elegant and
ysterious feeling, and facilitating the
alogue between human and nature.

ttled Tree and the Sound of
ence (Central Landscape Area)
e central garden takes the open
nshine lawn as the main space. It is
rounded by the garden space with
ferent characteristics, such as sparse
est grassland, terrace garden, mall
uare and sunken square. The organic
d network connects each space in
ies, combining the dynamic and
tic, and complements each other.
nse plants, natural wood, stones and
ulptures are used in the park, full of
e colors. Stone, wood, or implicit or
et, or zigzag or gentle, simple, direct
d subtle, so that the visitors can enjoy
surely and quietly.

问心湖（生态湖）

自然与景观和谐相融，让人随时可以感受到风的流连，聆听到树的呢喃，静听到云的低语，在这里生活便拥有湖水、绿地、林木、天空和流云，椰林挺秀，婀娜婆娑，水上闲庭，茶歇谜语，看庭前花开花落，望天空云卷云舒，不经意间洗浴疾劳，凝聚成心底最平淡、最诗意的感动。

水景崇尚自然，层次丰富，水面到地面过渡自然，平静的水映着天地、日月与星辰，在平静的空间中创造一种舒展式融入自然的体验。丰富的场景变化形成不同的空间体验，开敞的疏林草地、休憩活动空间、儿童活力园、健身步道穿行其中，空间氛围舒适，自然体验丰富。

◀ 阵列雕塑 Array Sculpture

enxin Lake (Ecological Lake)

e harmony between nature and landscape enables people to feel the wind at
y time, listen to the whisper of trees, listen to the whispers of clouds, and live here
th lake water, green land, trees, sky and flowing clouds. The coconut forest is tall
d clear, graceful and whirling. There are water courts, tea break riddles, flowers
oming and falling in front of the court, clouds rolling and comfortable in the sky,
thing and toiling carelessly.

aterscape loves nature with rich layers. The transition from water surface to
ound is natural. The calm water reflects the heaven and earth, the sun, the moon
d the stars, creating a kind of relaxing experience integrated into nature in the
aceful space. Rich scene changes forming different space experiences. Open
est and grassland, open space for leisure activities, children's energy park, and
ness footpath walk through them. The space atmosphere is comfortable and the
tural experience is rich.

顺德华侨城天鹅湖景观设计旨为配合项目建筑方案设计中所体现的地方特色和设计思想，设计追随 20 世纪最有影响力的建筑设计之一弗兰克赖特的流水别墅，讲究一致性，团结和谐和完整，这些设计理念改变了家的概念。我们吸取其中特点，采用天然材料，创造开阔视野，使自然融入室内环境，同时打破窗户内外景观之间的界限，并且采用最简单的景观要素，融入设计。

To display local characteristics and design ideas embodied in the project architectural design, the landscape design of Shunde OCT Swanlake closely follows the design style of Frank Wright's Falling Water, one of the most influential architectural designers in the 20th century, which emphasizes consistency, unity, harmony and integrity. In this regard, the concept of home has also been changed. We take full advantage of good design ideas, adopt natural materials to create a broad space, enable nature to access to the indoor environment while breaking the boundaries between interior and exterior landscape of the window, and make use of the simplest landscape elements in the design.

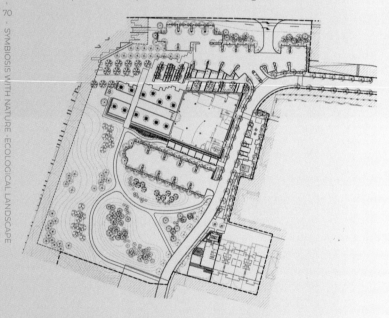

顺德华侨城天鹅湖
Shunde OCT Swanlake

设计时间：2014 年
项目委托：华侨城集团
项目地址：广东 佛山
项目规模：140000 ㎡

Design time: 2014
Entrusting party: OCT Group
Project site: Foshan, Guangdong
Project scale: 140, 000 ㎡

▲ 镜面水景 Mirror Water Feature

顺德天鹅湖以大自然流水雕琢的美丽形态为
设计初衷,保留顺德特色的岭南文化载体,整
体设计简约大气,还原自然。坐落于美丽的
桂畔湖畔的展示中心为顺德带来现代景观的
新表情。

Shunde Swanlake is designed with
the original intention that natural
water carves the shape of the lake, and
the lake retains the Lingnan cultural
heritage with Shunde characteristics.
The overall design is simple but
elegant, and at the same time the
nature is restored. Located at the
beautiful Guipan Lake, the exhibition
center brings a new expression of
modern landscape to Shunde.

▲ 景观跌水 Landscape Fa

▲ 亲水平台 Hydrophilic Platform

滨亭子 Lakeside Pavilion

◀ 林荫小道 Tree-lined Trail

▲ 休憩空间 Resting Space

互 动 景 观 - *S H A R E*

科技发展日新月异，生活中无处不智能，人对景观的需求不再止于纯粹的观赏，对景观的情感体验提出了更高的要求，"与人共享"的景观设计理念应运而生，互动景观突破传统景观的界限，增加景观对人类生活的功能性、艺术性、体验性等，形成以"景观"为核心的文化效应与情感体验。

Science and technology are changing with each passing day, and intelligence is available almost everywhere in our life. People's demand for landscape is no longer limited to pure viewing, instead, they put forward higher requirements for the emotional experience of landscape. The landscape design concept of "sharing with people" came into being. The interactive landscape breaks through the boundaries of traditional landscape, increasing the functionality, artistry and experience of the landscape to human life, and forming the cultural effect and emotional experience with "landscape" as its core.

大鹏新区具有天性独特与多样的地理资源，是包容着两种自然景观的
介面城区，让居民感受到自然与历史的共存。作品"海之韵"位于深
圳大鹏较场尾民宿小镇中轴线上，南侧面临大海与沙滩，是本次双年
展最接近海的作品之一。创作最初的想法就是"对话古今，植根鹏城
滨海历史文化"。作品通过模拟海浪外形的耐候钢雕塑创造出强烈的
视觉冲击，让观赏者在不同时间维度体会不同的光影变化，倾听海浪
的声音。装置加以"呼吸的沙"、"跳舞的沙"、"波光涟漪"三个
海元素互动装置与大鹏"历史"及"生态"产生对话，做出回响。

Dapeng New District has unique and diverse geographic
resources. It is an urban area that embraces two natural
landscapes, allowing residents to feel the coexistence of nature
and history. The work " Soul of the sea" is located on the
central axis of Jiaochangwei B & B Town, Dapeng New District,
Shenzhen. Facing the sea and the beach on the south side,
it is one of the works closest to the sea in this biennale. The
original idea of the creation is "talking between the ancient
and the modern, rooting in the coastal history and culture of
Pengcheng". The work creates a strong visual impact through
the weather-resistant steel sculpture that simulates the shape
of the waves, allowing viewers to experience different light and
shadow changes in different time dimensions, and listen to
the sound of the waves. The work interacts with the "history"
and "ecology" of the Dapeng New District through three
interactive installations: "Breathing Sand", "Dancing Sand", and
"Silver Water".

2019 深港城市 / 建筑双城双年展参展作品
2019 Shenzhen-Hong Kong Bi-city Biennale of Urbanism / Architecture

海之韵 / 城脉·律动
Soul of the Sea / Bridge-City Pulse · Rhythm

设计时间: 2019 年
项目地址: 广东 深圳
项目规模: 20m x 15m x 3.8m （长 x 宽 x 高）
项目类型: 参与型互动装置雕塑
材料: 耐候钢

Design time: 2019
Project site: Guangdong, Shenzhen
Project size: 20m x 15m x 3.8m (L x W x H)
Project type: Participatory interactive installation sculpture
Material: Weather-resistant steel

作品"海之韵"视觉向海岸线开放，进而让此区域成为观察自然变化的容器。因此设计的焦点并非建立一个新构筑物，而是创造关联起海洋与小镇的艺术介质。项目原场地是个由冰冷的混凝土地面构成的闲置广场，"海之韵"装置重新定义地面与海的介面，通过科技、人文艺术的融合，创造出一个与自然互动或对话的全新舞台。

互动装置"呼吸的沙"最初的灵感源自设计师儿时海边抓螃蟹的记忆。海浪冲刷过沙滩，身形小巧的螃蟹在白色细沙下呼吸，沙面轻微起伏冒泡，留下了小小的印记。为了还原记忆，设计师通过艺术加工，让沙子以生命呼吸的韵律进行动态起伏。足下之沙，仿佛产生了流动的生命，让参与者在不经意中感受到大自然的魅力，同时引发对海洋环境的思考。

The visual of the work " Soul of the sea " is open to the coastline, which makes this area a container for observing natural changes. Therefore, the focus of the design is not to build a new structure, but to create an artistic medium that connects the ocean and the town. The original site of the project is an idle plaza made of cold concrete. " Soul of the sea " redefines the interface between the ground and the sea, and creates a new stage of interaction or dialogue with nature through the fusion of technology, humanities and art.

The original inspiration for the interactive installation "Breathing Sand" comes from the designer's childhood memory of catching crabs by the sea. After the waves wash the beach, the tiny crabs breath under the fine white sand, and the sand surface bounces and bubbles slightly, leaving a small mark. In order to restore the memory, the designer uses artistic processing to make the sand fluctuate dynamically with the rhythm of breathing. The sand underfoot seems to have life, let the participants inadvertently feel the charm of nature, and at the same time trigger the thinking of the marine environment.

▲ 海之韵模型效果图
Soul of the Sea Model Renderings

▼ 装置轮廓演变过程
Device Outline Evolution Process

① ② ③

④ ⑤ ⑥

▲ 海之韵细节图
Soul of the Sea Detail

▲ 海之韵细节图
Soul of the Sea Detail

▲ 俯瞰图 Overlook

▲ 夜景 Night Scene

"跳舞的沙"与"呼吸的沙"呼应而诞生、一张一翕、一快一慢的节奏，将儿时海边玩耍的氛围推到高潮。设计师用夸张的手法，巧妙地赋海沙以"沙舞者"的身份，我们通过程序控制，将沙舞者的舞姿与音乐韵律结合，让观赏者感受在地热温泉喷发状态的沙所呈现的沙浪翻滚、此起彼伏的特殊形态，创造出舞与乐的意境场景。

轻踏的水面荡起层层涟漪，这是童年戏水的画面，"波光涟漪"灯光互动装置由此而诞生。

以沙面作为载体，人们的活动与灯面产生交互作用，产生了多种光波跳跃的互动感应。变幻的灯光创造出由内而外层层扩散的水波效果，犹如水面泛起的层层波浪，产生特有的趣味性，让人回想起童年戏水的欢乐时光，那似乎又是对历史的一种提醒与回顾。

The interactive device "Dancing Sand" echoes the "Breathing Sand". Open and then close, fast and then slow, driving the atmosphere of childhood play by the sea to the climax. The designer exaggeratedly gives the sand on the beach the identity of "sand dancer". Through program control, we can combine the dance pose of the sand dancer with the rhythm of music, so that the viewers can feel the rolling of sand appearing in the eruption state of the geothermal hot spring, creating a artistic conception of dance and music.

After gently stepping over, the water surface ripples, which is the picture of our childhood playing in the water. The "Silver Water" light interactive device is born then.

With the sand surface as a carrier, people's activities interact with the surface of the lights, producing interactions of a variety of light waves. The changing light creates a water wave effect that spreads from the inside to the outside, just like the waves that rise on the surface of the water, creating a unique fun, and at the same time recalling the happy time of childhood playing in the water, which seems to be a reminder and review of history.

▲ 海之韵细节图
Soul of the Sea Detail

在整个设计过程中，设计团队对此主题曾经有三个方案进行诠释，我们尝试多维多角度去观察和思考，从视觉、听觉以及触觉的感受出发，在设计思路上探索新的转变，经过了多轮设计讨论，内部专家评选，深圳区政府及策展公司的综合评定后，最终选定"海之韵"方案为我司的参展方案。

"海之韵"设计突显了工业空间的特色：采用耐候钢为整个装置的造型主元素，外形轮廓模拟海浪冲刷形态。未经修饰的耐候钢龙骨，犹如航行百年，船身经历海浪侵蚀后残留下的船架，斑驳锈迹，具有浓郁的历史色彩。龙骨的汇集型线条引导观众进入装置内部，耐候钢与海沙形成了一个粗糙又统整的美感，历史与生态的元素相互穿插，呼应大鹏的海滩印象，同时形成了极具文化氛围的展示空间。

Throughout the design process, the design team has proposed three proposals on this topic. We try to observe and think in multiple dimensions and angles, start from the perception of sight, hearing and touch, and explore new changes in design ideas. After many rounds of design discussions, internal expert selection, and comprehensive evaluation by the Shenzhen Municipal Government and the curatorial company, the " Soul of the sea " program was finally selected as our company's participation program.

The design of " Soul of the sea " highlights the characteristics of industrial space: Weathering steel is used as the main element of the whole device, and the shape and contour simulate the shape scoured by sea waves. The undecorated weather-resistant steel keel is like a ship frame that has been eroded by the waves after sailing for hundred years. The mottled rust has a rich historical color. The combined lines of the keel guide the audience into the installation. Weather-resistant steel and sea sand form a rough and unified beauty. The elements of history and ecology intersect each other, echoing the beach impression of Dapeng New District, and at the same time forming a very cultural atmosphere of exhibition space.

桥 - 城脉·律动

该桥梁设计位于深圳大鹏所城与较场尾之间，大鹏所城南门中轴线上，鹏飞路南侧。该桥东西走向，跨度约四十多米，河水清澈，河道两侧自然生态迷人，可在桥的东西两侧观看日出日落。

海浪冲击沙滩形成的自然波浪曲线，提取其自然的曲线作为造桥元素，自然围合出东西两侧观景平台。游客沉浸于落日余晖的映照中，情不自禁地停留其间，或坐或站于两侧看台，这时的桥已不单单是供人通行的交通工具，其本身就是一道景观。

曲线解决了桥与道路、广场的高差，并将大鹏古城的中轴线与对岸的较场尾的广场上的海之韵相结合，同时运用曲线的元素使之成为整体，并形成了一升一降、一阴一阳的符合中国传统文化中道法自然的传统理念，完美解读了阴阳在自然中的和谐融合。

使用现代科技增强趣味性，两侧观景台可随季节变化随之升降，与生态景观密切贴合。丰水期升起，枯水期下降，贴近水面，使游客沉浸其间。并在桥下方设置数字水幕和体感音乐喷泉，游客可通过手机操作，控制水幕投影出不同的图案、文字等；也可在特定感应区做出不同行为，产生不同的声光效果。材质上选用传统的石笼做法与海边贝壳相结合，形成独特的贝壳笼。

空间关系设计对比
Comparison of Spatial Relation Design

一般桥梁 General Bridge　　设计桥梁 Design Bridge

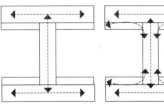

公共空间与桥梁成垂直关系，河流沿岸与桥梁联系不紧密
The public space is perpendicular to the bridge, and the river bank is not closely connected to the bridge

公共空间与桥梁联通融合，且河流沿岸与桥梁联系紧密
The public space is connected to the bridge, and the river bank is closely connected to the bridge

CSC 20th SELECTED PROJECT

▲ 城脉 · 律动效果图
City Pulse-Rhythm Effect Picture

Bridge-City Pulse-Rhythm

The designed bridge is located between Peng City and Jiao Changwei in Shenzhen, on the central axis of the south gate of Peng City and on the south side of Pengfei Road. The bridge spans over forty meters from east to west. The natural ecology on both sides of the clear river is charming and you can watch sunrise and sunset on both sides of the bridge.

The natural wave curve formed by the sea waves hitting the beach is extracted as a bridge-building element to naturally enclose the viewing platforms on the east and west sides. Tourists are immersed in the sunset and can't help staying here, sitting or standing on the viewing platforms on both sides. At this moment, the bridge is not only a means of transportation for people to pass through but also a landscape in itself.

The curve solves difference in height between bridge and the road and the square, and combines the central axis of ancient Peng City and the rhyme of sea on the Jiao Changwei square on the other side. It makes a whole by using the element of curve, and forms the traditional concept of "rise and fall, yin and yang" which conforms to the traditional Chinese culture of Tao following nature, perfectly interpreting the harmonious integration of yin and yang in nature.

Using modern science and technology to enhance enjoyment, the viewing platforms on both sides can rise and fall with seasonal changes and closely fit with the ecological landscape. They rise in the rain season and fall in the dry season, keeping close to the water surface to allow tourists to immerse themselves in the beautiful scenery. A digital water curtain and a somatosensory music fountain are set up under the bridge. Tourists can operate and control the water curtain to project different patterns and characters through their mobile phones. They can also make different behaviors in specific induction zone to produce different acousto-optic effects. The traditional gabion procedure is combined with seaside shells to form a unique shell cage.

▲ 城脉·律动模型图 Bridge-City Pulse-Rhythm Mo

▶ 城脉·律动效果图 Bridge-City Pulse-Rhythm Eff

平面图 Plan

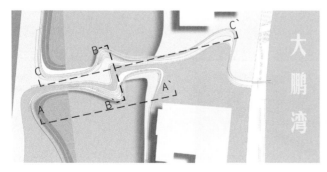

大
鹏
湾

A-A' 立面图 Elevation A-A'

C-C' 剖面图 Profile C-C'　　　　　　　　　　　　　　　B-B' 剖面图 Profile B-B'

赛瑞景观二十周年作品特辑

我们探寻真正的场景精神，

强调人与场地的互动及感知，

这远不止于眼前的浮华或装饰的富奢。

采用人性化的设计，自然的语言，

阐述新城市生活模式，

这是我们对这个项目最初的想法。

We explore the true spirit of the scene,

emphasizing the interaction and

perception between people and the

venue.

This is far more than the luxuriance of

the eyes or the luxury of decoration.

Adopt humanized design, natural

language,

explain the lifestyle of the new city,

this is our initial thought of this project.

郑州海马青风公园
Zhengzhou Haima Qingfeng Park

设计时间：2016 年	Design time: 2016
项目委托：海马地产	Entrusting party: Haima Property
项目地址：河南 郑州	Project site: Zhengzhou, Henan
项目规模：7669 ㎡	Project scale: 7, 669 ㎡

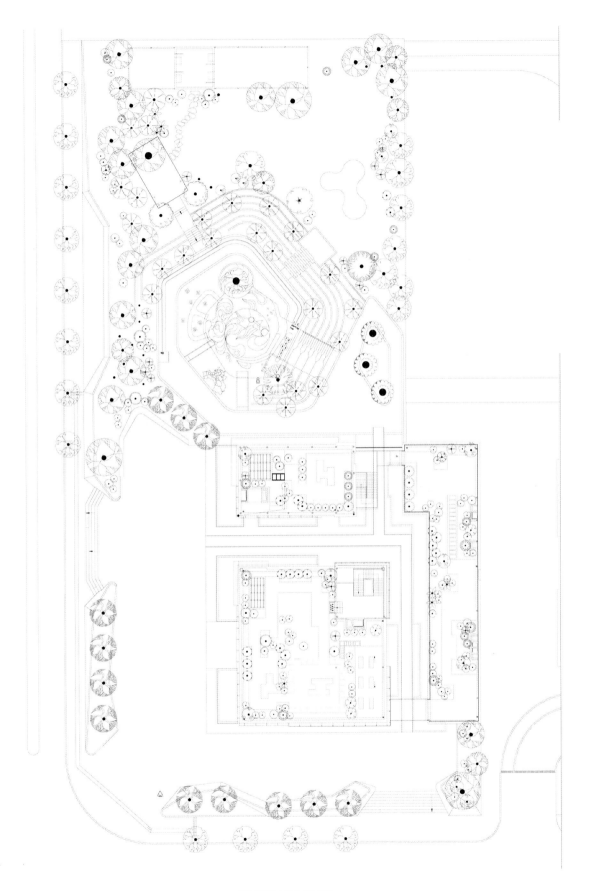

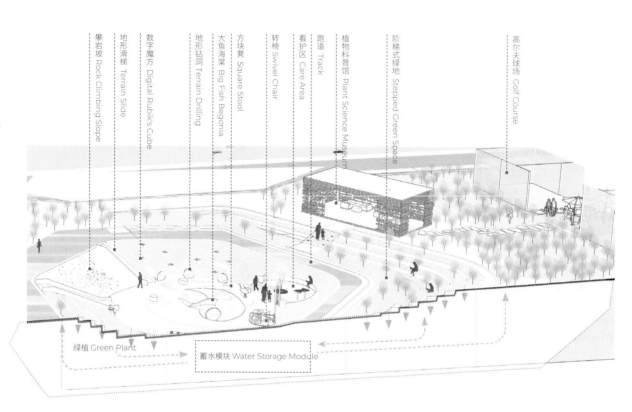

攀岩坡 Rock Climbing Slope
地形滑梯 Terrain Slide
数字魔方 Digital Rubik's Cube
地形钻洞 Terrain Drilling
大鱼海棠 Big Fish Begonia
方块凳 Square Stool
转椅 Swivel Chair
看护区 Care Area
跑道 Track
植物科普馆 Plant Science Museum
阶梯式绿地 Stepped Green Space
高尔夫球场 Golf Course

绿植 Green Plant

蓄水模块 Water Storage Module

▲ 生态植物科普园 Ecological Plant Science Park

▲ 展示区俯视图 Display Area Top View

▲ 下沉儿童乐园 Sinking Children's Playground

目位于郑州市经开第十八大街，占地面积 50843m²，整个项目定
绿色""环保""节能"，也是开发商"青风"系列项目的首开
。展示区遵循场地属性，从设计本真出发，探索自然本源，回归
文化。这是我们的设计初衷。利用场地本身洼地，营造儿童成长
认知空间，结合雨洪管理系统，利用生态节能环保材料。营造具
文关怀、自然生态的儿童成长空间。场地的原有洼地具备了营造
体验空间的先决条件，因此我们从：

型高尔夫·山

谷童音·谷

物科普·林

区域去满足儿童的不同需求和玩法，释放儿童天性。抓住儿童最
的滑、爬、钻、跑等玩乐方式。利用螺旋形的跑道、滑梯、钻洞，
识文，以打击乐等不同形式的设施，搭建小朋友相互之间交流的
。

ted at Jingkai 18th Avenue in Zhengzhou, this project
rs an area of 50,843m². The project, aiming to be "green",
ronmentally friendly" and "energy-saving", is the first of
gfeng" series of the developer. Demonstration area, based
he fundamental idea of designing and corresponding
e features of the site, searches the origin of nature and
ns to the local culture. This is the original intention of the
gn. In our design, we turn the low-lying land into the space
hildren to explore and gain knowledge. We also adopt the
n water management system and environmental-friendly
erials for ecological and energy-saving purpose. The
e with humanistic care and natural ecology is formed for
ren's growth. Since the low-lying land is the precondition
tablish a distinguished experience space, with the
ving areas:

cro-golf · Mountain

hildren's Music · Valley

ant Science · Forest

下沉儿童乐园 Sinking Children's Playground

过对植物的认知，从植物生长到开花结果，感悟自然植物的生命，结合雨洪管理系统，了解植物的生命期。 植物科普和自然发现相结合，寓教于乐。

个场地从儿童玩乐、植物认知、自然运动体验等不同方面，结合场地现有的特征，设计适合场地属性、文关怀的儿童体验空间，诠释了人性化、生态性、环保低价的设计理念。真正意义上把"绿色"、"生"、"节能"、"环保"诠释到整个项目中。

重场地，从场景出发，顺势而为

调功能的体现，反对装饰化的景观设计，利用现场洼地，营造不同形式的儿童玩乐空间。搭建儿童相互

流的桥梁，释放童真天性。

态环保

合雨洪管理系统，采用生态节能材料。场地形成一个自我循环的小气候，减少材料浪费和造价。

色科普

计儿童植物科普认知园，感悟自然生命，探索自然的亲子体验之旅。

We create three spaces to meet various needs and requirements of children to release their true nature. Based on children's ways of playing, such as sliding, climbing, drilling, running, etc, we use the spiral runway, slide, holes for crawling, characters and pictures, percussion musical instruments and other forms of facilities to build a bridge for children to communicate with each other.

Learning detailed knowledge about plants, from seeds to their flowers and fruits, children can have a deep understanding about their lives and life cycles with the storm water management system. By combining plant science and nature exploration, children can learn about nature happily.

The features of the site are designed on its existing characteristics for different purposes, including playing, plant knowledge learning and natural sports experience. The children's experience space with humanistic care shows the design ideas of humanity, ecology, environment protection and low cost. The principles of being "green", "ecological", "energy-saving" and "environmentally friendly" have been integrated into the whole project.

Respect the site and seek the most appropriate design, go with the natural terrain

The design attaches great importance to magnifying functions rather than artificial landscape for decoration purpose. By using the existing low-lying land, we create a different play space for children. The design focuses on building the bond for children to communicate with each other and releasing their true nature.

Eco-environmental Protection

Ecological and energy-saving materials are used of with the storm water management system. A self-cycling climate system is developed on the site, and material waste and building costs are reduced.

Green Science Popularization

Build plant science garden for children to gain an understanding on other lives on the planet. It is a trip of nature exploration for a whole family.

① Surface Runoff
② Plant
③ Infiltration

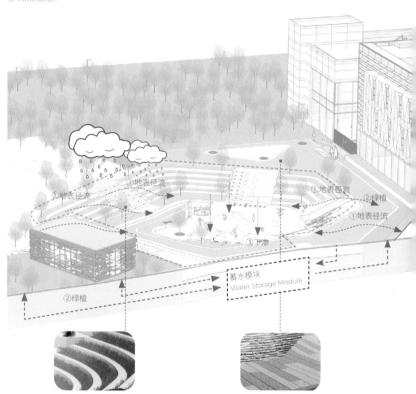

①地表径流
②绿植
③下渗
蓄水模块
Water Storage Module

清水混凝土
清水混凝土表面不需要装饰材料，较为清洁和环保
Fair-faced Concrete
The surface of fair-faced concrete does not need decorative materials and is relatively clean and environmental

生态石
可以杜绝白蚁和霉菌，具有良好的生态效益
Ecological Stone
It can eradicate termites and mold, with a good ecological benefit

▲ 下沉儿童乐园 Sinking Children's Playground

▲ 植物科普园 Plant Science Park

项目位于有着莲城之称的湘潭。民间流传："万楼兴，湘潭兴"。经历了四个世纪的风雨，万楼以其雄伟滂沱的气势、深厚凝重的文化积淀，成为湘潭繁荣兴旺的时代象征和湘潭人民的精神依归。利用万楼带来的第一门户机遇，联结莲城文脉，强化万楼片区的新城形象。联系万楼与本案，述说从古到今的时代过渡。

The project is located in Xiangtan, known as "Lotus City". After four centuries of wind and rain, Wanlou, with its magnificent momentum and profound cultural accumulation, has become the symbol of the times of prosperity and the spiritual support of Xiangtan people. Taking advantage of the first portal opportunity brought by Wanlou, connecting the context of Liancheng, and strengthening the image of new city in Wanlou district, we build the bridge between the ancient times to the present.

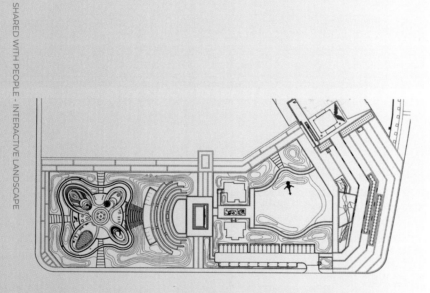

湘潭长房·万楼公馆
Chanfine · Wanlou Mansion in Xiangtan City

设计时间：2019 年	Design time: 2019
项目委托：长房集团	Entrusting party: CHANFINE Group
项目地址：湖南 湘潭	Project site: Xiangtan, Hunan
项目规模：17928 ㎡	Project scale: 17, 928 ㎡

时空之门 The Gate of Time and Space

文脉传承 - 百年书院

每座城市都有铭刻文化记忆的建筑，对湘潭而言，万楼书院便是这样的存在。承袭四百年万楼文化精神，藏古纳今，又以长房·万楼公馆营销中心的身份伫立湘江之滨。用现代元素和符号进行演绎，建筑远观如书，以开放的姿态迎接各方来者，展现着一座城的胸怀及建设者的抱负。它与时代美学共鸣，焕新城市居住理想，为湖湘名仕筑就一城仰望的府邸。

Cultural Heritage - Centennial Academy

There are buildings with cultural memory in every city. For Xiangtan, Wanlou Academy is such an existence. Inheriting the cultural spirit of Wanlou buildings for four million years, collecting the past and embracing the present, and standing by the Xiangjiang River as the marketing center of Longfang Wanlou mansion. The modern elements and symbols are used for interpretation. It welcomes all the visitors with an open attitude and shows the city's and builder's ambition. It resonates with the aesthetics of the times, rejuvenates the city's ideal of living, and builds a mansion for the famous officials in Huxiang.

◀ 万楼书院
Wanlou Academy

▼ 万楼时空广场
Wanlou Space-time Plaza

创意互动高科技水幕 - 时空之门

时空之门的出现,点燃了整个城市广场,对话万楼,与时空对话,引导人们和城市空间发生互动。时空之门既是雕塑,也是标识,具有独特性识别性。

圆形数控水幕呈现时,绚丽的灯光,仿佛穿梭的时空隧道,通过水中汀步,穿梭在其中,带来丰富的视觉、听觉、触觉等多方面的空间感知体验,提升场地的价值与记忆点。圆形数控水幕停止时,呈现的是天空与门的融合倒影,镜面反射着云朵风云变幻的色彩,表达时空之门与万楼的浓浓历史感的相互呼应。

互动景观的关键不只是人与设施的互动,更重要的是提供人与人的互动交流的场所,最终形成高标识的湘潭门户。

Creative Interactive High-tech Water Curtain - The Gate of Time and Space

The appearance of the door of time and space ignites the whole city square, which talks with thousands of buildings and guides people to interact with the city space. The gate of time and space is not only a sculpture, but also a sign, with unique identification.

When the circular numerical control water curtain is presented, the brilliant lights, like the shuttle space-time tunnel, shuttle in it through the water step, bringing rich visual, auditory, tactile and other aspects of space perception experience, improving the value and memory of the site. When the circular numerical control water curtain stops, it presents the fusion reflection of the sky and the door. The mirror reflects the changeable color of the clouds, and expresses the mutual echo of the time and space door and the thick sense of history of Wanlou.

The key to the interactive landscape is not only the interaction between people and facilities, but also the provision of a place for people to interact and exchange, and ultimately the formation of a Xiangtan portal.

SHARED WITH PEOPLE - INTERACTIVE LANDSCAPE

▶ 时空之门详图 Time and Space Door Detail

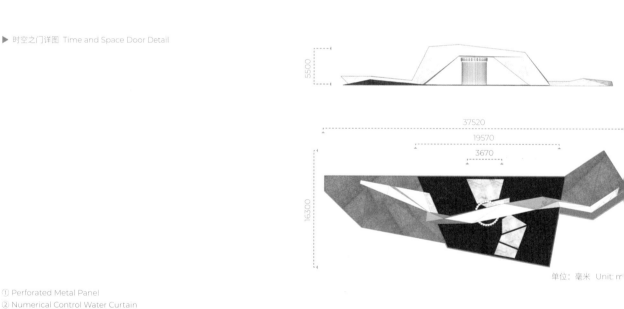

单位: 毫米 Unit: m

① Perforated Metal Panel
② Numerical Control Water Curtain
③ Scattered 70mm Black Gravel
④ 600 × 600 ×30(THK)Glossy Black Galaxy
⑤ 600 × 600 ×30(THK)Water Rinse Sand

① 金属穿孑
② 数控水
③ 散置 70 厚黑色砟
④ 600×600×30(厚) 光面黑色
⑤ 600×600×30(厚) 水洗面浪流

▲ 时空之门 The Gate of Time and Space

▲ 时空之门 The Gate of Time and Space

公共参与 - 万楼音乐剧场

万楼音乐剧场，是户外多功能活动场地。让浮躁干涩的信息世界浸润在诗意的温柔中。闲暇的周末与家人看一场户外电影，或与朋友一起野炊。这里将作为未来公共活动的共享空间，通过场地将莲城的居民汇聚于此，在喧嚣的城市里，造一处静心修怀之所。

无限趣玩 - 万楼儿童乐园

万楼儿童乐园提取莲花形态，新奇的视觉体验，打造立体主题公园。利用起伏变化的地形，设计滑梯、攀岩。让孩子们在上面打滚、攀爬、俯冲、滑行、躲藏而令他们感到惊奇。儿童游乐、户外跑道、有氧共享，这是一个全龄的共享活动空间。

blic Participation - Wanlou Music Theater

anlou music theater is an outdoor multi-functional activity venue. Let the petuous and dry information world infiltrate in the poetic tenderness. Watch an tdoor movie with your family or have a picnic with your friends on the weekends. a shared space for public activities in the future, it will gather the residents of ncheng through the venue and create a place for meditation in the noisy city.

inite Fun - Wanlou Children's Paradise

anlou children's park extracts lotus form and novel visual experience to create nree-dimensional theme park. Using the undulating terrain, design slide and k climbing, design enable children to roll, climb, dive, slide and hide. Including ldren's entertainment, outdoor track, aerobic area, this is a full-age shared ivity space.

▶ 万楼儿童乐园
Wanlou Children's Paradise

万楼兴街

利用万楼带来的第一门户机遇，联结莲城文脉，强化万楼片区的新城形象。

设计亮点分为：

1. 主题互动 —— 激发活力

万楼主题 IP 水景

主题灯光装置

互动跳跃灯光装置

2. 文化活动 —— 复兴文艺

不定时文化集市

企业宣传活动

不定时的艺术装置

3. 多样参与 —— 社交纽带

户外主题美食街

户外全龄健身活动

户外音乐剧场

4. 绿色生活 —— 健康生态

绝美花海 / 回忆

户外健康活力场所

户外健身仓

▲ 艺术互动广场 Art Interactive Plaza

▲ 艺术互动剧场 Art Interactive Theatre

Wanlou Xing Street

Taking advantage of the first portal
opportunity brought by Wanlou, our
project links the context of Liancheng
to strengthen the new image of
Wanlou district.

Highlights of the Design are as Follows:

1. Theme Interaction: Stimulating
Vitality

Wanlou theme IP waterscape

Theme lighting device

Interactive jump lighting device

2. Cultural Activities: Reviving Literature
and Art

Irregular cultural fair

Corporate publicity activity

Irregular art installation

3. Diverse Participation: Social Bond

Outdoor theme food street

Outdoor full-age fitness activity

Outdoor music theatre

4. Green Life: Healthy Ecology

Beautiful sea full of flowers / memory

Outdoor healthy and vigorous site

Outdoor fitness room

▲ 艺术灯光互动广场 Art Light Interactive Plaza

▲ 湘潭名人讲堂 Xiangtan Hall of Fame

茶与文化
与文化

主 题 景 观 - *EMPATHY*

在地域特色趋同问题日益显现的今天，我们究竟需要什么样的城市，
承载什么样的生活空间，成为都市更新与城市可持续发展的重要议题。
"与文化共情" 景观设计理念，从地域文化、场地特点、项目定位等
方面，赋予脉络，创造情景，重组价值，把现有的地域和场地特色，集
合成城市文化可持续发展的环境。

With the convergence of regional characteristics increasingly
becoming obvious, what kind of city we need and what kind
of living space we carry have become the major issues of city
renewal and urban sustainable development. The landscape
design concept of "empathy with culture" endows context,
creates scenarios, and reorganizes value from the aspects of
regional culture, site characteristic and project positioning,
integrating the existing region and site characteristics into the
environment of sustainable development of urban culture.

湖南省博物馆荟萃了湖湘大地的文物遗珍，展现了湘楚文明的来龙去脉，本案意在采用现代风格的景观设计来突显湖南省博物馆的战略位置，使其成为宣传湖湘文化的最佳窗口。

Hunan Museum gathers relics and treasures all over the province, which display the profound history of this land. This project intends to highlight the strategic position of Hunan Museum using modern landscape designs, and make it the best window to showcase and publicize Human cultures.

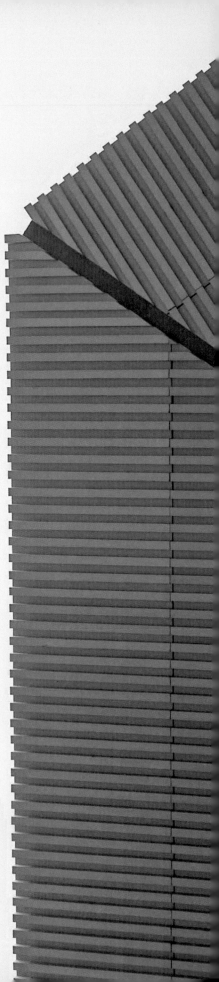

湖南省博物馆
Hunan Museum

设计时间：2012 年
项目委托：湖南省博物馆
项目地址：湖南 长沙
项目规模：25653 ㎡

Design time: 2012
Entrusting party: Hunan Museum
Project site: Changsha, Hunan
Project scale: 25, 653 ㎡

湖南省博物馆是湖南省最大的综合性历史艺术博物馆，也是全国优秀的爱国主义教育示范基地和湖南省 AAAA 级旅游景点之一。

基于现代风格的定位，因此延续附近公园里比较生态自然风格的树，方案设计成成行成组的树阵，通过平衡引力和特质，创造比较开阔的视野空间来清晰地突显临街的博物馆，同时与建筑主入口中心轴线平衡。对比博物馆设计方案，结合该博物馆基地现状，使用现代、沉默、开放的草坪和隆起的缓坡，在不同的标高面上使用树阵和一些灌木，使树木看起来更加有秩序，种植区可让建筑轮廓更加清晰显眼，更具现代风格，使博物馆战略位置更加突出。

The Hunan Museum is the largest comprehensive historical art museum in Hunan Province. It is also the national outstanding patriotism education demonstration base and one of the AAAA-level tourist attractions in Hunan Province. Based on modern style, instead of arranging trees in natural style as in the nearby parks,the design arranges trees into rows and groups. By balancing gravity and traits, it creates a relatively wide field of vision to clearly highlight the museum facing the street while balancing the central axis of the building's main entrance. Based on the museum base's conditions, using modern, silent and open lawn and gently used slopes are put in place. It uses tree arrays and some shrubs on different elevations to make the trees look more orderly. The planting area makes the building's outline clearer and more modern, making the strategic location of the museum more prominent.

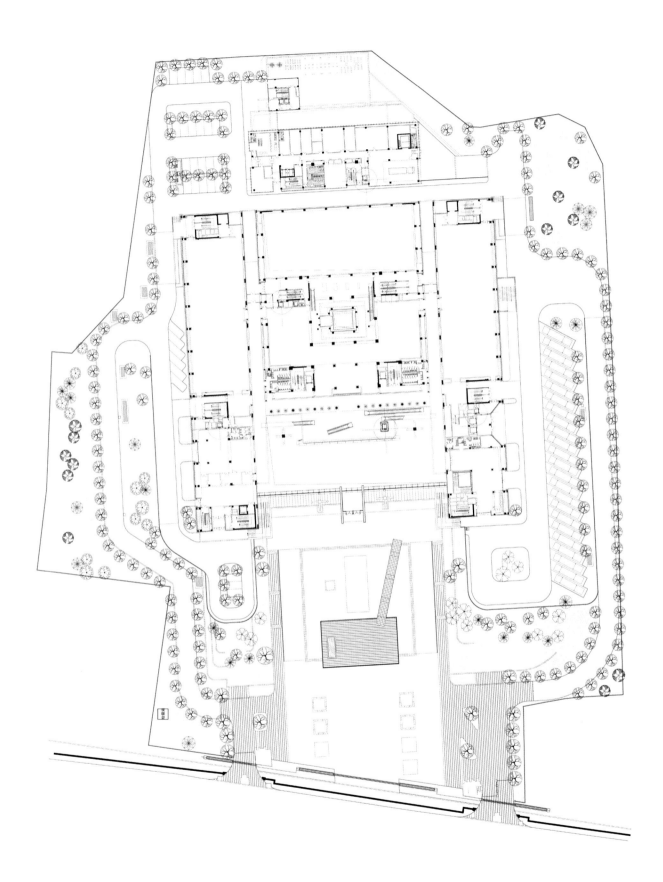

鼎盛洞庭

建筑以"鼎盛洞庭"为创作源泉，从鼎的意象、气势及文化精神内涵入手，转化为湖南省博物馆的建筑外形，同时，顶部造型为水形结晶体，象征三湘四水，彰显湖湘文化深厚的底蕴。整体为对称式布局，彰显其庄重、沉稳的风格特征。

Prosperous Dongting

The building is inspired by"Dingsheng Dongting", starting from Ding's imagery, momentum and cultural spirit, transforming into the architectural shape of the Hunan Museum. At the same time, the top shape is water-shaped crystal, symbolizing the "Sanxiang Sishui", highlighting the profound foundation of Hunan culture. The overall layout is a symmetrical layout, highlighting its solemn and calm style.

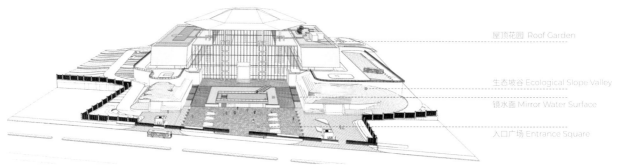

屋顶花园 Roof Garden

生态坡谷 Ecological Slope Valley

镜水面 Mirror Water Surface

入口广场 Entrance Square

与文化共情 · 王府井大饭店

▲ 大门 Gate

景观设计中，使用均衡组合的手法，来强调轴线的景观节点。通过良好的比例构成，实现建筑与景观的完美平衡。

In landscape design, balanced combination is used to emphasize the node of the axis. The perfect balance between architecture and landscape can be achieved through good proportion.

▲ 大门 Gate

迎宾广场冷暖灰色调铺装图案，来自中国五行元素中的火和水，通过变化和组合结合建筑概念。

休息区域和树池，同样也是中国五行元素中的一种。使用竹子和木座椅的组团，将"木"元素植入该区。

Welcome square uses warm and cool grey to pave patterns, which comes from the fire and water in China's Five Elements, combining architectural concepts through change and composition.

The resting area and the tree pool are also elements of China's Five Elements. The cluster uses bamboo and wooden chairs to add "wood" into the mix.

▶ 镜面水池 Mirror Pool

▼ 特色灌木种植 Characteristic Shrub Planting

随着城市的发展，荷叶塘、烟竹塘消失了，我
们一直在思考，独特的景观，它所要具备的
因素是什么？它吸引人群的亮点是什么？而
人们梦想中的家又是什么？所以设计以追寻
逝去的记忆、延续城市的文脉为主旨，新中
式风格的定位，寻找留存的书院文明，立意
于中国山水写意的意境，结合西方的造园手
法，创造出东意西境的园林景观。

Lotus ponds and Yanzhu ponds have
disappeared with the development of
cities. We have been thinking, what
are the necessary factors for a unique
landscape? What makes it attractive to
people? What does everyone's dream
house look like? Therefore, our design
intends to pursue the past memory
and extend the culture of the city.
The keynote of the design is the new
Chinese style, in order to search for
the remaining academic civilization.
The design aims at combining artistic
conception of Chinese landscape
painting with the western gardening
techniques, in order to create a
landscape of both Chinese and western
art.

长房潭房·时代公馆展示区
Chanfine Tan Fang · Time Mansion Exhibition Area

设计时间：2019 年
项目委托：长房集团
项目地址：湖南 湘潭
项目规模：8187 ㎡

Design time: 2019
Entrusting party: Chanfine Group
Project site: Xiangtan, Hunan
Project scale: 8, 187 ㎡

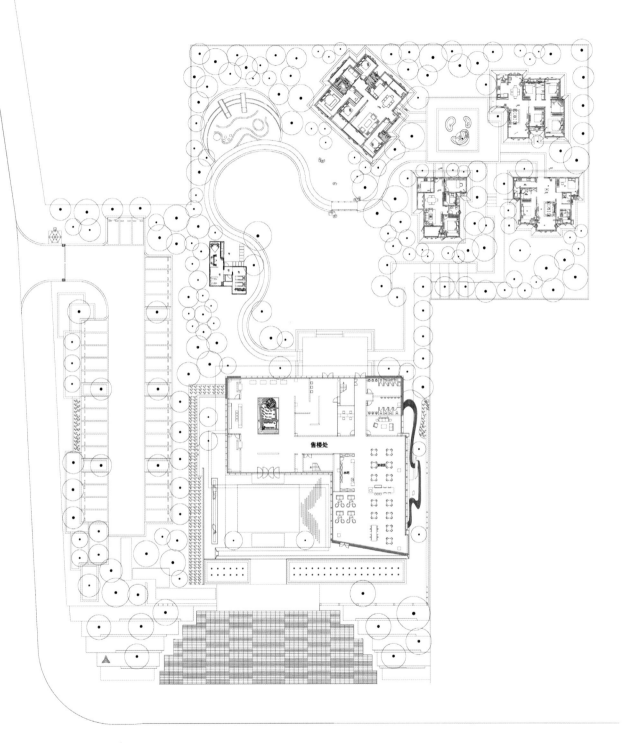

售楼处

长房潭房时代公馆位于湖南省湘潭市岳塘区。整体设计分为"礼仪入口"、"山水竹林"、"花溪谷"三大空间。"迎"、"停"、"赏"、"憩"、"享"五大主题路线。

迎：利落的手法，创书香雅韵，迎四方之客，造宾至如归。

停：停看山水竹林，品味湖湘茶韵。

赏：绿坡花谷，赏花海之韵。运动健身，享健康之美。

憩：过一池三山，憩精品华宅。

享：享童趣乐园，听欢声笑语。

Chanfine Tan Fang· Times Mansion is located in Yuetang District, Xiangtan City, Hunan Province.

The overall design is divided into three major space: "etiquette entrance", "landscape bamboo" and "Huaxi valley" . Five theme routes of "welcome", "stop", "appreciate", "rest" and "enjoy".

Welcome: create the elegant charm of books, welcome the guests from all places and make guests feel at home.

Stop: stop to look at the landscape of bamboo, and taste the charm of Hunan tea.

Appreciate: green hills and flower valley, appreciate the charm of flower sea and exercise to enjoy the beauty of health.

Rest: go through one pond and three mountains, and have a rest in a boutique house.

Enjoy: enjoy children's paradise, listen to their laughter.

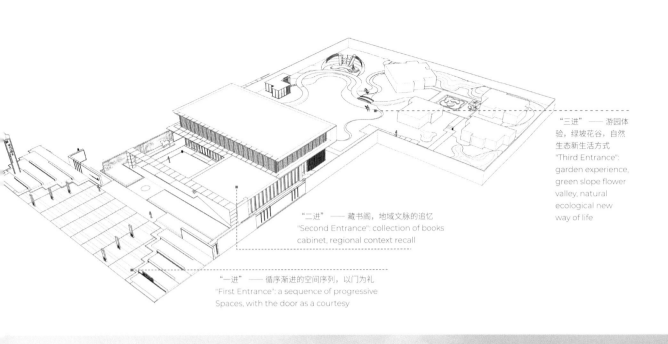

"三进" —— 游园体验，绿坡花谷，自然生态新生活方式
"Third Entrance": garden experience, green slope flower valley, natural ecological new way of life

"二进" —— 藏书阁，地域文脉的追忆
"Second Entrance": collection of books cabinet, regional context recall

"一进" —— 循序渐进的空间序列，以门为礼
"First Entrance": a sequence of progressive Spaces, with the door as a courtesy

▲ 书香广场 Shuxiang Square

延续城市的文脉

中国古代书院是儒学思想的传播基地，因此，传统儒学"礼"、"仁"、"乐"的思想内核决定了书院教学活动的主要内容包括三个方面：祭祀行礼、躬行践履、悠游山林，相对应地，中国古代书院的功能性质主要包括以下三方面内容：

1. 礼仪场所 - 孔庙
2. 治学场所 - 讲堂、御书楼
3. 游息场所 - 书院园林

Continue the Context of the City

The ancient Chinese academies were the bases for the dissemination of Confucianism. Therefore, the core of traditional Confucian thoughts of "ceremony," "benevolent", and "music" determined that the main contents of the academy's teaching activities consisted of three aspects: sacrifices, practice, and excursion in the mountains and forests. Correspondingly, the functional nature of ancient Chinese academies mainly includes the following three aspects:

1. Etiquette place-Confucian Temple
2. School place-Lecture hall, Imperial Book Building
3. Resting place-Academy Garden

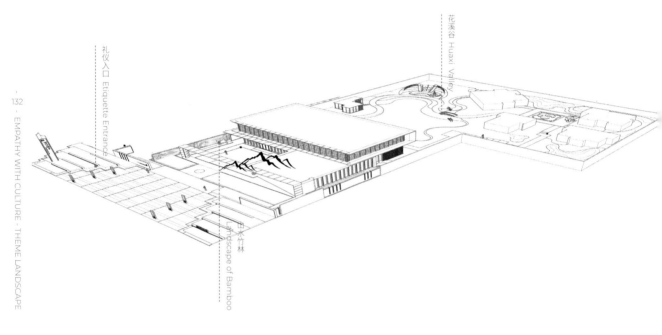

礼仪入口 Etiquette Entrance

山水竹林 Landscape of Bamboo

花溪谷 Huaxi Valley

▲ 时代公馆展示区空间布局 Space Layout of Time Mansion Exhibition Area

礼仪入口

精神堡垒以竹简为形，创书香雅韵，迎四方之客，造宾至如归。

山水竹林

运用现代手法，通过对山、竹、光影的设计组合，创造出山水竹林的禅意空间。

花溪谷

以自然肌理诠释现代风格，步入花溪，看疏林草坡，欣赏光影摇曳、绿坡花溪的景观效果。

Etiquette Entrance

The spirit fortress is shaped by bamboo slips, creating a fragrant and elegant charm, welcoming guests from all directions, and making guests feel at home.

Landscape of Bamboo

Using modern techniques, through the design and combination of mountains, bamboo, light and shadow, to create a zen-like space of landscaped bamboo.

Huaxi Valley

Interpret the modern style with natural textures, step into Huaxi, watch the sparse forest and grass slopes, and enjoy the swaying light and green slopes of Huaxi's landscape effects.

▲ 映雪堂 Yingxuetang

▲ 山水竹林 Landscape of Bamboo

大门入口设计大面积镜面水景,配合涌泉,种植高干疏枝的植物于层叠的景墙旁边,烘托配合建筑,营造一湖一境的深幽氛围。入口中庭用现代手法,通过对山、竹、光影的设计组合,创造出山水竹林的禅意空间。后场通过对山谷、坡地、花草等自然元素的构造,人路过花影桥,步入花溪,看疏林草坡,欣赏这光影摇曳、绿坡花溪的景观效果。在样板房的中心景观带,通过对园路的规划以及功能的追求,设计出一池三山的景观空间。

A large mirror surface waterscape is designed at the entrance, and matches up with fountain. High-trunk and sparse-branches plants are planted next to landscape wall, to match up with buildings and create a serene scene. The atrium of the entrance adopts modern technique, to create space of Buddhist mood through the combination of mountain, bamboo and light and shadow. For the backyard, with the construction of natural elements such as valley, slope, flowers and plants, people can walk through flower bridge and cross over flower stream, look at the open forest and grass slope, and appreciate the view of swaying light and shadow, green slope and flower stream. In the central scenery zone of the model house, a landscape of one pond and three mountains is designed, with planning of roads and pursuing functions.

▲ 镜面水景 Mirror Surface Waterscape

时代公馆项目以湘潭书院为文化背景,依山水竹林再现场地记忆,营造了一处静心素雅、花溪坡谷的诗意空间。项目以循序渐变的线条、玻璃、水墨、竹林、绢画等现代设计手法传承场地文脉和记忆,通过空间的渐进,展现湘学始源的博厚悠远,应时而借一日四时之景,描绘出一幅烟竹浩渺、白云悠悠的山水画面。

The Times Mansion Project uses the Xiangtan Academy as the cultural background, recreates the site memory with the landscape of bamboo, and creates a quiet and elegant poetic space in Huaxi Po Valley. The project inherits the context and memory of the site with modern design methods such as progressively changing lines, glass, ink, bamboo, and silk painting. Through the gradual progress of the space, it displays the richness and longevity of the origin of Hunan Studies.By showing the scenery of four different stages of one day, it depicts a vast landscape of bamboo and white clouds.

▶ 中庭 Courtya

《长物志》中说，居山水间者为上。寻山问水，仿佛是文人雅士天然的追求和理想。然浮华都市中，隐秘尚是奢谈，何以山水之境？故永安建发·玺院展示区清晖园从"居然山水之间，情寄世外燕城"的理念开始。以留园为蓝本、以传统造园原则来演绎国人对"以隐逸为高，以游放山水为傲"的文人情怀。一步步塑造古代文人雅士的"居尘而出尘"的江南山水园林。

"Long Things" said that the people living in the mountains and waters are better. Looking for mountains and asking for water is like the pursuit and ideal of the literati. In the glitz of the city, the secret is still a luxury talk, why is it the landscape? Therefore, Yongan Jianfa feels that the Qinghui Garden in the exhibition area of the brothel begins with the concept of "between the mountains and rivers, and sentiment to the outside world". Taking the Lingering Garden as the blueprint and using the principle of traditional gardening to interpret the literati feelings of "the people who are high in seclusion and proud of the mountains and rivers". Step by step to shape the Jiangnan landscape garden of the ancient literati.

永安建发·玺院展示区
Yongan Jianfa · Xiyuan Exhibition Area

设计时间：2017 年
项目委托：建发地产
项目地址：福建 永安
项目规模：3435 ㎡

Design time: 2017
Entrusting party: C&D Real Estate
Project site: Yongan, Fujian
Project scale: 3, 435 ㎡

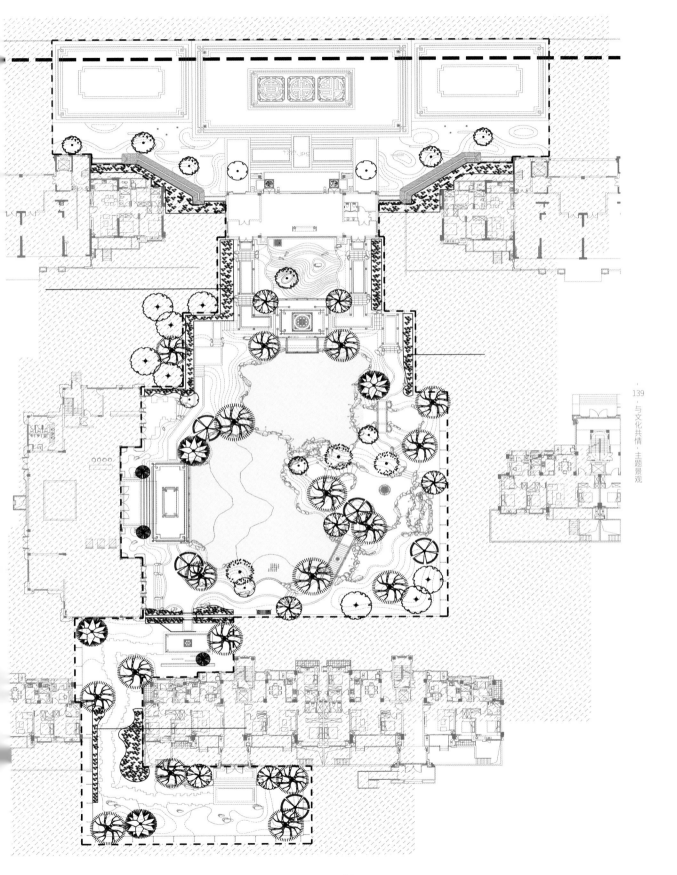

整个示范区采用的是王府礼制，三进院的进门礼仪，以三进院落，言说山水间的归家礼序。三进院是：一进乾元庭，二进"四水归堂"，三进涵碧园。每一进院营造了不同的意境与氛围。一进院高门阔府，万千威仪；二进泗水归堂，棋定天下；三进院师法江南，山水意蕴。

打造中式园林特点：在情感上符合传统中国文化礼制，在功能上遵循现代人的生活习惯特征，在空间序列礼仪上，乾元庭与泗水归堂为华文唐风中式景观，涵碧园为仿古中式自然山水院落空间。涵碧园重点营造景点为框景造画——溪山清远图，一幅步步生景的山水画卷。

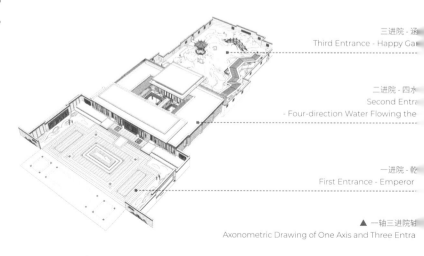

三进院 - 涵
Third Entrance - Happy Gar

二进院 - 四水
Second Entra
- Four-direction Water Flowing the

一进院 - 乾
First Entrance - Emperor

▲ 一轴三进院轴
Axonometric Drawing of One Axis and Three Entra

▲ 乾元庭 · Emperor Yard

e whole demonstration area adopts the ceremony system of Wang House and entrance ceremony of these three entrances,

ing that the returning home-coming ceremony between mountains and rivers.

ese three entrances include: first entrance - Emperor Yard, second entrance - "Four-direction Water Flowing the Hall", third

trance - Happy Garden. Each entrance creates the different artistic conception and atmosphere, first entrance - magnificent

ace, mighty and powerful; second entrance - four-direction water flowing the hall, dominant and controllable; third entrance

arning from nature, displaying the connotation between mountains and rivers.

eate the garden featured with Chinese characteristics: emotionally, it should conform to the traditional Chinese cultural

remony system; functionally, it should abide by the living habit characteristics of modern people; spatially (sequence

quette), Emperor Yard and Four-direction Water Flowing the Hall belong to Chinese landscape with Tang style, and

ppy Garden serves as an archaistic Chinese natural landscape courtyard. The representative made by Happy Garden is the

framed scenery as for scenic spots: Qingshan Drawing of Xishan, a scroll of landscape painting.

一进院 - 乾元庭

打造高门阔府，再现尊贵气派的礼序仪式。高墙深院，曲径通幽，俯仰之间，帘幕无穷数。构建沉稳色彩体系，延展大美中式礼仪，延续了唐代建筑出檐深远的屋檐风格，呈现"如鸟斯革，如翚斯飞"的飞檐形态，追求力与美的统一。

二进院 - 四水归堂

四水归堂是安徽江南民居独有的平面布局方式，其建筑风格理念来源于徽州文化传承天人合一的理学思想。四水归堂式住宅的个体建筑以传统的"间"为基本单元，各单体建筑之间以廊相连，和院墙一起，围成封闭式院落。屋顶内侧坡的雨水从四面流入天井，民间称之为肥水不流外人田。

First Entrance - "Emperor Yard"

Create a high gate and wide house, reproduce the noble style of ritual ceremony. High walls and deep courtyards, winding paths pass quietly, between pitching, the curtain is infinite. Establish a calm color system and extend Chinese etiquette. Continuing the eaves style of far-reaching eaves of the Tang Dynasty architecture, presents the eaves form of "like a bird's leather, like a flying bird", pursuing the unit of power and beauty.

Second Entrance - "Four-direction Water Flowing the Hall"

"Four-direction Water Flowing the Hall" is a unique plane layout of Jiangnan dwellings in Anhui Province, whose architectural style is derived from the natural science of Huizhou culture of "man is an integral part of nature". Built on the basis of traditional basic unit of "room", individual buildings of "Four-direction Water Flowing the Hall" are connected by corridors, thus forming a closed courtyard with the courtyard wall. Since rain on the inside slope of the roof flows into the courtyard from four directions, it was called "every miller draws water to his own mill."

▶ 乾元庭 Emperor Y

▼ 山水景墙 Landscape W

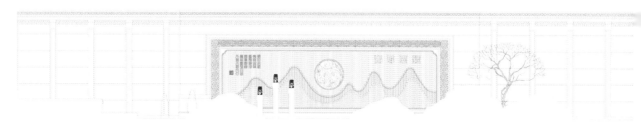

三进院 - 涵碧园

三进涵碧园第一个景点为展示区的核心景点框景造画——溪山清远图，景点名取自南宋四大家夏圭的传世佳作《溪山清远图》。以留园为蓝本、以现代手法来演绎国人对"以隐逸为高，以游放山水为傲"的文人精神的追求。

造园原则：

1. 构架山水

2. 模拟仙境

3. 移天缩地

4. 诗情画意 融入其中

三曲廊

廊在建筑历史中，一直都是皇家和贵族的专属，是古代皇室们休闲雅集的重要场所。曲廊依墙又离墙，因而在廊与墙之间组成各式小院，曲折有法。或在其间栽花置石，或略添小景而成曲廊。

揽月桥

揽月桥以石材雕刻拼成，拱如人间彩虹，两侧垂柳包含着欣欣向荣、"春常在"的美好祝愿，桥身与水面倒影形成一轮圆月，象征美满团圆之意。桥面上有牡丹地花，寓意长寿富贵、寿比南山。行至桥中最高处远望，欣赏清风池之景。水瀑流下清风池，太湖石布置于池周，有山影重重、重峦叠嶂之姿。

▼ 摘星亭 Zhaixing Pavilion

Third entrance - "Happy Garden"

The first attraction of third entrance is the frame painting of the core attractions of the exhibition area: Xishan Qingyuan Map. The name of the scenic spot is taken from the hand-picked masterpiece of the Northern Song Dynasty, Xia Gui, "Xishan Qingyuan Map". Taking the Lingering Garden as the blueprint and using modern methods to interpret the pursuit of the literati spirit of "the people who are high in seclusion and proud of the mountains and rivers".

Gardening principle:

1. Landscape-based structure

2. Simulated wonderland

3. Miniaturization

4. Poetic illusion and integration

▼ 水中禅院 Water Zen Temple

nqu Gallery

the history of architecture, the corridor, always being the exclusive symbol of royal families and nobilities, is an important place r the ancient royal family to relax. Since the winding corridor leans against and separates from the wall, various small courtyards e therefore developed between the corridor and the wall in a roundabout way. Or plant flowers and place stones in the middle, grow small plants in the corridor.

nyue Bridge

nyue Bridge is carved with stones and arched like rainbows on the earth. Its weeping willows on both sides representing st wishes of flourish and "always being the spring" and water reflection form the full moon, which symbolizes happiness and union. There are peonies on the bridge, meaning longevity and wealth. Looking into the distance at the highest point in the idge, enjoy viewing the breeze pool. The waterfall flows past the breeze pool, and Taihu Stone is arranged around the pool sed with unnumbered mountain shadows and peaks.

▲ 涵碧园 Happy Garden

依托建筑的设计理念及风格，本项目景观设计以"月满星城 山水潇湘"为设计理念，通过"中国 - 湖湘 - 长沙"的本土语言向世界发声，向世人展示出一个充满湖湘传统韵味、具有鲜活生命力的国际化山水洲城新形象，形成一个具有高度识别性的高端国际会议中心，与潇湘长沙对话，与世界对话。

Relying on the design concept and style of the building, the landscape design of the project takes "the star city with mountains and lakes" as the design concept, and uses the local language of Changsha to speak to the world, shows the world a new image of an international landscape city full of the traditional charm of Hunan and has fresh vitality, forming a high-end international conference center with a high degree of identification, bridging Xiaoxiang Changsha with the world.

长沙国际会议中心
Changsha International Conference Center

设计时间：2019 年
项目委托：长沙环球世纪发展有限公司
项目地址：湖南 长沙
项目规模：146632 ㎡

Design time: 2019
Entrusting party: Changsha Global Century Development Co., Ltd.
Project site: Changsha, Hunan
Project scale: 146, 632 ㎡

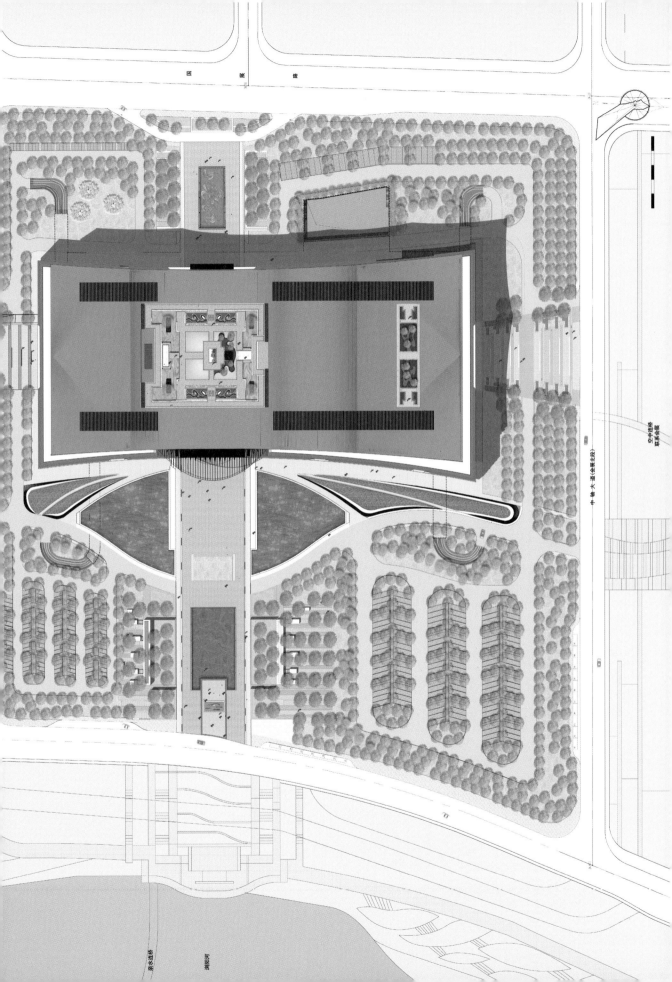

展览中心位于长沙城市群发展核心区、未来城市副中心。西侧为浏阳河，东侧为湿地公园，南侧为城市绿化轴。

我们从片区整体山水格局思考，以国际会议中心"山"为主体烘托对象，以浏阳河"水"为水云，以磨盘洲为"洲"，以会展新城为"城"。将其打造成一副山水画意、翰墨中华的山水画卷，向世界展示中国形象。

The exhibition center is located in the core area of Changsha urban agglomeratio development district and the sub center of the future city, Liuyang River in the west, wetland park in the East and urban green axis in the south.

Considering the overall landscape pattern of the area, we take the "mountain" the International Conference Center as the main object, the "water" of Liuyang River a the water cloud, the Mopanzhou as the "continent", and exhibition new city as the "city" to create a calligraphy of China's mountains and lakes to show the world.

国际会议中心
onal Conference Center

西入口广场
West Entrance Plaza

会议中心布局
Conference Center Layout

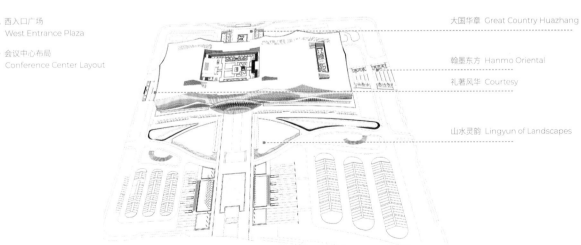

大国华章 Great Country Huazhang

翰墨东方 Hanmo Oriental

礼著风华 Courtesy

山水灵韵 Lingyun of Landscapes

▲ 西入口广场夜景
Night View of West Entrance Plaza

◀ VIP 东入口水景
VIP East Entrance Water Feature

入口空间

山水灵韵： 西入口水景倒映建筑形态，与建筑月门相互呼应、相辅相成，形成阴晴圆缺的景观变化，营造情景交融之美，以"山水灵韵"的意境开篇。

大国华章： 东入口运用中国代表性的文化符号 —— 祥云，打造"大国华章"的尺度呼应。

翰墨东方： 南入口应用"书卷"的元素，加入中非地刻，丰富铺装形式，传递着"翰墨东方"的文化内涵。

礼著风华： 北入口利用传统"框景"的手法，延续南入口"书卷"理念，从南至北如一幅徐徐展开的书卷，打造"礼著风华"的空间感受。

Entrance Space

Landscape Aura: The water view of the west entrance reflects the architectural form, which echoes and complements each other with the moon gate of the building, forming a cloudy and sunny landscape change, creating the beauty of the combination of feelings and scenes, and starting with the artistic conception of "landscape aura".

Country Spirit: Auspicious cloud, a representative cultural symbol of China, is used in the east entrance to echo the scale of a big country spirit.

Oriental Culture: The south entrance uses the elements of "book scroll", adds carvings, enriches the pavement forms, and conveys the cultural connotation of "oriental culture".

Etiquette and Elegance: The north entrance uses the traditional "frame view" method to continue the concept of "book scroll" in the south entrance. From the south to the north, it is like a slowly unfolding book to create a spacial experience of "courtesy, etiquette and elegance".

入口 LOGO 标识设计

过三维结构，营造量子山水雕塑 LOGO，雕线条硬朗且色彩统一鲜明。同时根据观察角的不同营造 360° 不同的视觉盛宴。

go Design of Main Entrance

ough the three-dimensional ucture, the quantum landscape lpture logo is created, the lines he sculpture are strong and the ours are uniform and bright. At the ne time, according to the different wing angles, 360 degree visual feast nsured.

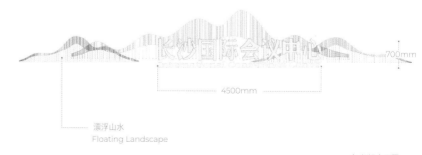

漂浮山水
Floating Landscape

700mm

4500mm

▲ 标识立面图
Logo Elevation

中轴仪式感 - 三重礼序

延承岳麓书院空间布局，中心空间呈"凸"字形结构。

Axis ritual sense-triple etiquette

Continuing the space layout of Yuelu Academy, the central
space presents the "convex" shape.

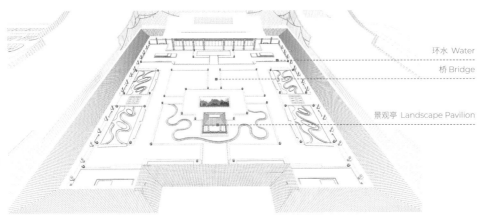

環水 Water

桥 Bridge

景观亭 Landscape Pavilion

▲ 会议中心空间布局
Conference Center Space Layout

核心功能空间：屋顶花园等

以岳麓书院为切入点，延承岳麓书院的中轴
对称布局与纵深多进的院落形式形成，均质
体量、平行并置的空间序列，体现书院礼序
和湖湘印记。

Core Functional Space: Roof Garden, etc.

With Yuelu Academy as the starting
point, the space sequence of
homogeneous volume and parallel
juxtaposition is formed by extending
the axial symmetry layout of Yuelu
Academy, courtyards are built with
more depth, reflecting the ritual
sequence of academy and the legacy
of Hunan.

▲ 屋顶花园竹林
Roof Garden Bamboo Forest

项目以"源于西方，碰撞民国，融于梦幻芙蓉里"作为展示中心的设计理念，整体打造出体现法式花园的轴线感与秩序感。设计中通过轴线对称式布局呈现出从简单到复杂的各种空间表现形式。入口营造仪式感、阵列感，中心草坪汇聚人气、烘托气氛，花影亭打造诗意文化空间，树阵广场点缀芙蓉花水钵，打造特色记忆点。记忆中的民国时期的长沙就从这些情景中缓缓浮现。

The design philosophy of this project is "origin from the West, inspired by Republic of China and blending into the dreamy hibiscus", which shows the sense of axes and order of French gardens. The axisymmetric layout reveals various space forms from simple to complex ones. The entrance creates a sense of ceremony and array, central lawn gathers popularity and drives atmosphere, flower pavilion builds poetic culture space and tree array embellishes hibiscus water bowel, creating special memory points. Memory about Changsha in the Republic of China period emerges from these scenes.

长沙环球世纪·未来城展示区
Exhibition Area of Changsha Global Century · Future City

设计时间：2019 年
项目委托：长沙环球世纪发展有限公司
项目地址：湖南 长沙
项目规模：8888 ㎡

Design time: 2019
Entrusting party: Changsha Global Century Development Co., Ltd.
Project site: Changsha, Hunan
Project scale: 8, 888 ㎡

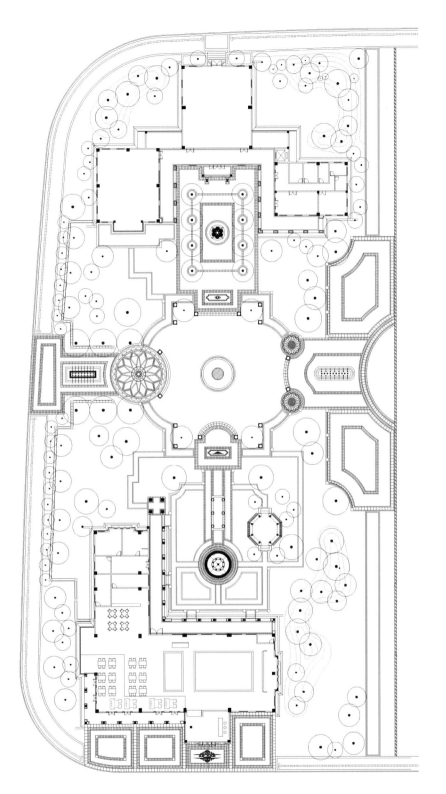

▲ 树阵广场 Tree Plaza

环球世纪·未来城以民国风格为
计基调,用东魂西技的手法延续
国时期精神,将传统的结构形式
过重新设计组合,以另一种民族
色的标志符号再出现 。我们在
湘重遇这个充满魅力的时代,用
朵芙蓉, 致敬那旖旎风情。

▲ 酒店前广场 Hotel Front Square

▲ 中心草坪 Central Lawn

Global Century Future City , following the style of the Republic of China as the keynote, and taking full advantage of Western technologies combined with the Eastern spirits, re-designs and re-combines the traditional structural form to reappear with another national symbol. We experience this glamorous era again in Hunan province, and pay a tribute to the charming style with a lotus.

树阵广场俯瞰 Tree Plaza Overlooking
花影亭 Flower Booth

佛山是岭南文化的源头，我们在这座文化底蕴深厚的城园中，传承岭南文化的诗意生活，让禅文化融入景观中，将园林空间画境升华到意境，让有限的自然山水艺术提供审美体验的无限可能。

Foshan is the source of Lingnan culture. We inherit the poetic life of Lingnan culture, integrate Zen culture into the landscape, sublimate the landscape space to the artistic conception, and let the limited natural landscape art provide the infinite possibility of aesthetic experience in this city park with profound cultural connotation.

佛山龙光玖龙郡
Acesite Fegion in Foshan

设计时间：2019 年
项目委托：龙光地产
项目地址：广东 佛山
项目规模：6646 ㎡

Design time: 2019
Entrusting party: LOGAN Real Estate
Project site: Foshan, Guangdong
Project scale: 6, 646 ㎡

▲ 登门望府 Dengmen Wangfu

相采用折线与曲线结合的设计语言打造尊

···雅、活力有序的现代中式都市花园。

···异景间，国风自现，三重院落六种心境，

···序自然。

···门为礼、庭为和、墙为序、水为镜、院为

···的院落空间布局，打造新中式院落景观。

···采用大尺度的延展面，用简洁精致的模

···化金属，构建富有韵律的有序排列，展现

···开放、自信的国际设计风范。细节设计保

···方风骨的安之若素。安宁朴素的自然美

···在绿植与石材的碰撞下激发新的感官，呈

···出空间中细微的感动，笔墨深浅，寂寥无

··· 勾勒出悠闲的山水意境。

···s project combines broken line

··· d curve in design to build exalted,

··· gant, vigorous and well-organized

··· odern Chinese city park.

··· epping into the spectacle room,

··· ental style comes into sight. Three

··· ferent yards bring six different

··· oods, ritual sequence well organized.

··· e entrance adopts a large-scale

··· tension surface, simple and exquisite

··· odular metal, to build a rhythmic and

··· derly arrangement, showing an open

··· d confident international design

··· le. The details keep the Oriental

··· le. The peaceful and simple natural

··· auty stimulates new senses under

··· e collision of green plants and stone

··· aterials, presenting the subtle touch

··· the space, the ink deep and silent,

··· tlining the leisurely landscape mood.

▲ 枯山水 Dry Mountain Water

◀ 迎宾门楼 Welcome Gate

本案以北大文化为设计蓝本，延续王府的雍容华贵和沉稳雅致的特有气质，运用中式园林的手法处理多变空间，延续中式及建筑新中式主义元素，营造能触动视觉和强化心灵归宿感的景观，打造适合现代人生活的人文气息和健康社区是本设计的宗旨。

项目秉承"文化设计"的理念，深入挖掘北大文化内涵，并与地域和人文相融合。将新中式元素与北大文化相结合，以现代人的审美需求打造富有韵味的景观，让北大文化在现代景观表现形式中完美地体现出来，让使用者感受到传统文化与现代设计的相融合。

景观设计上以人为本，充分考虑当地人生活习俗与需求来设计。结合现有的条件，创造更加丰富的空间与造型。在设计形式上采用集中打造，重点打造，以平衡造价，提高品质。

The case based on the design of Peking University Culture, continuing the palace's graceful and elegant and unique temperament, using Chinese gardens to deal with the changing space, continuing the Chinese and new Chinese elements of architecture, to create a landscape that can touch the vision and strengthen the soul. It is the purpose of this design to create a humane atmosphere and a healthy community suitable for modern people's lives.

Adhering to the concept of "cultural design", the project taps into the cultural connotation of Peking University and integrates with the region and humanities. Combining the new Chinese elements with the Peking University culture, creating a rich landscape with the aesthetic needs of modern people, the Peking University culture is perfectly reflected in the modern landscape expression, allowing users to experience the integration of traditional culture and modern design.

The landscape design is people-oriented, and it is designed with full consideration of the customs and needs of local people. Combine existing conditions to create richer spaces and shapes. In the design form, we will focus on creating and focusing on building to balance the cost and improve the quality.

广州北大资源博雅 1898

Guangzhou PKU Resources Boya 1898

设计时间：2016 年
项目委托：北大资源集团
项目地址：广东 广州
项目规模：24369 ㎡

Design time: 2016
Entrusting party: PKU RESOURCES
Project site: Guangzhou, Guangdong
Project scale: 24, 369 ㎡

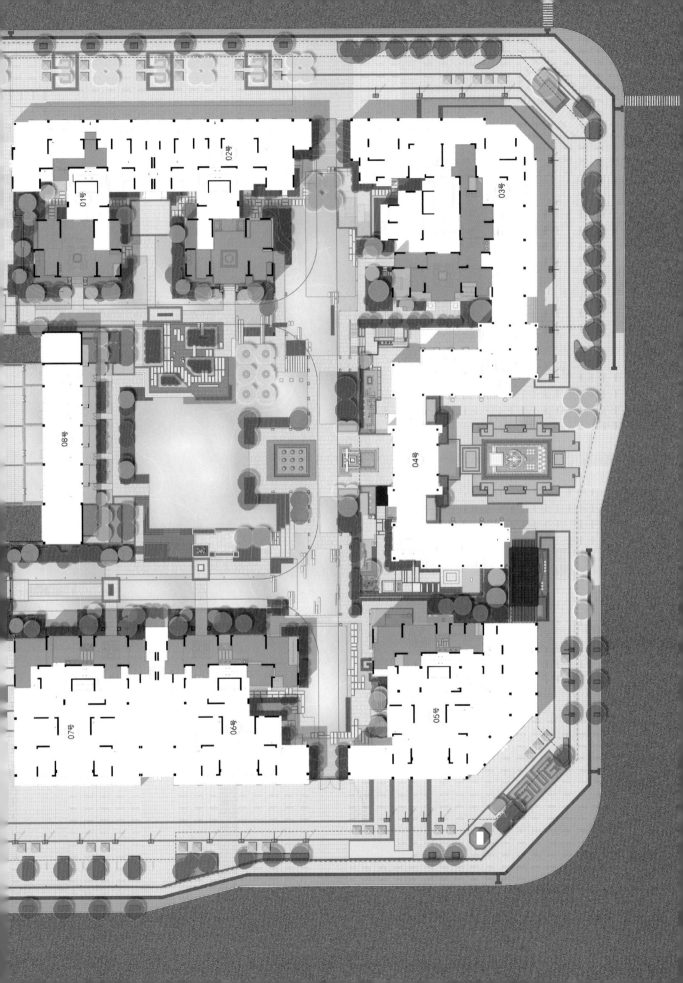

景观设计之灵感源自燕园（北京大学）文化，融合广府人居文化之精髓，打通南北景观效果，点缀北大园内景观元素，将源自北大未名湖中的"翻尾石鱼"元素、方正字库与南方的鱼图腾文化进行融合延伸，以铺装雕刻、铁艺、景墙等作为载体，实现景观与人文的结合，使居住者找到一种融入感和认同感。项目承接北大文化，融合南方理念，潜心研究中国南北造园艺术，用现代设计手法演绎传统文化，遵循以人为本的原则，充分考虑当地人生活习俗与需求来设计，提供更舒适、更人性化的生活空间。

The landscape is designed upon the inspiration from the culture of Yanyuan (the Peking University). On one hand, it combines the essence of settlement cultures of Guangfu people, breaks through the effects of the north-south landscape, and embellishes the landscape elements in the Peking University. On the other hand, integrates the elements of the "Roll-Tailed Stone Fish" from the Peking University Weiming Lake, the FounderType, and the fish totem culture of the south, and reflects both the landscape and humanity with paving sculptures, iron arts, and landscape walls as carriers, enabling residents to find a sense of integration and identity.

Carrying on the culture of Peking University, and integrating the concepts of the south, the project centers on studying the art of gardening in the north and south of China to deduce the traditional culture with modern design methods, follows the principle of "human - oriented", and takes into full consideration of the local customs and needs, so as to provide more comfortable and humane living space

▲ 营销中心 Marketing Center

▲ 营销中心 Marketing Center

迎宾小雕塑
Welcome Small Sculpture

石趣休憩小园
Stone Grand with Stone Sitting & Special Tree

对于古人所追求的"结庐在人境，而无车马喧"的生活理想，以及悠游山水的怡然情怀，我们现代人亦趋之。

建瓯玺院项目在遵循"一轴三进院"的传统礼序空间之上，巧妙地融入当地茶文化，以茶为引，结合江南园林造景手法，以紧凑的布局和别有洞天的诗意山水院落，在山水之间营造出了一处别具禅茶之韵、闲情隐逸之风的居住场所。

For the ideal life pursued by ancients as described by poetry of "In people's haunt I build my cot; Of wheel's and hoof's noise I hear not" and their pleasant mood of immersing in natural landscape, we modern people desire as well.

Abiding by the traditional ritual sequence of "one central axes and three courtyards",guiding by the tea,combining with the landscape design techniques of Jiangnan gardens, building with compact structure and particular poetic courtyard, Jianou Royal Land project integrates the local tea culture to construct a residential place among mountains and rivers with Zen and tea charm and hidden leisure.

建瓯建发玺院
Jianou Jianfa Royal Land

设计时间：2018 年
项目委托：建发地产
项目地址：福建 建瓯
项目规模：4690 ㎡

Design time: 2018
Entrusting party: C&D Real Estate
Project site: Jianou, Fujian
Project scale: 4, 690 ㎡

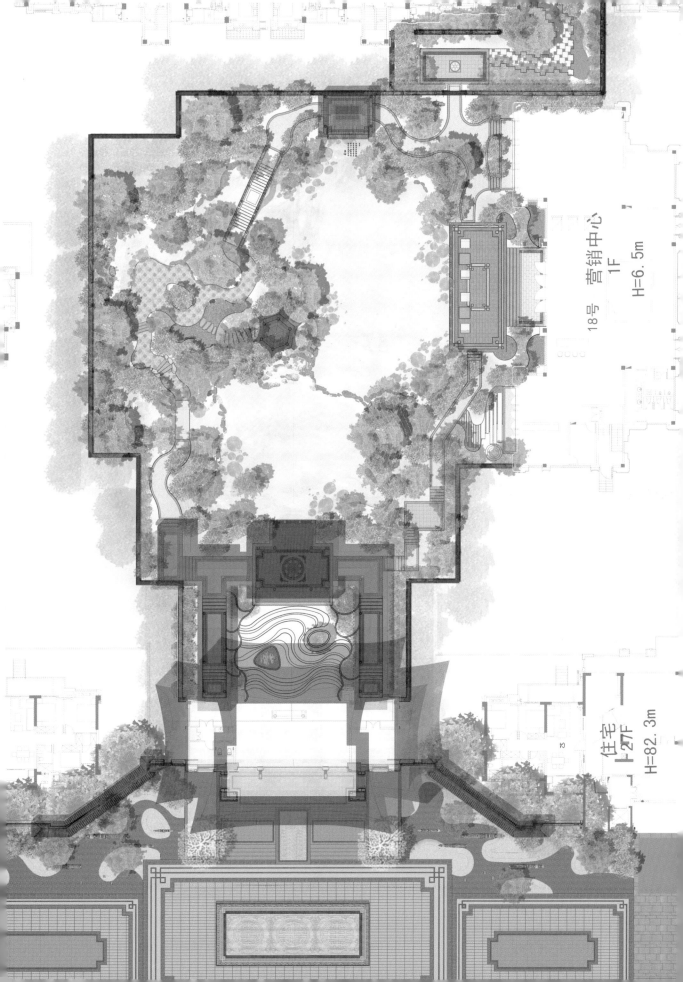

18号 营销中心
1F
H=6.5m

住宅
27F
H=82.3m

▲ 禅茶之院 Zen Tea House

项目以"东方茶韵，水云天境"的居住理念，
将空间的变化与禅茶文化相契合，在发扬江
南园林造园精神的同时，也传承了当地的文
脉底蕴，使人在游历山水时亦可感悟"禅茶
一味，天人合一"的哲学思想。

基于优越的地理条件，设计以自然诗意栖居
和东方文化传承为起点，院落空间的塑造和
传统空间场所的精神营造成为此次设计的重
点和亮点。

崇文庭 Chongwen Yard

袖云门 Cloud Gate

▲ 禅茶之院 Zen Tea House   冷香亭 Lengxiang Pavilion   ▲ 崇文庭 Chongwen Yard

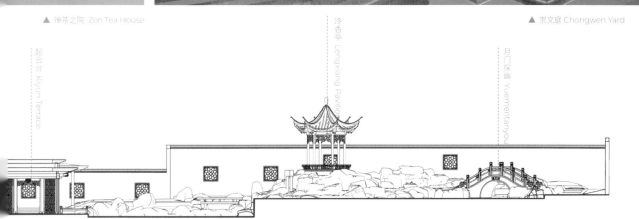

起云台 Kiyun Terrace   月门探幽 Yuementianyou

▲ 立面图 Elevation

设计从最讲究的"门面"开始，以"五开三启门"的规制为统领，采用古代大门中等级最高、空间最阔的"王府大门形制"，两侧影壁八字排开，五十五米的府门威严纵横。这种尺度的大门，在古时非王公贵胄之宅不可采用，是主人显赫地位的象征。古之礼制，以门彰显。两侧影壁能够升聚院内人气，而且有助于钱财的积聚。大门则采用九宫钉大门，九宫钉大门体现皇家最高规格礼制。

禅茶之院采用四水归堂的布局，四周均为回廊。四水归堂是安徽江南民居独有的平面布局方式，其建筑风格理念来源于徽州文化传承天人合一的理学思想。四水归堂式住宅的个体建筑以传统的"间"为基本单元，各单体建筑之间以廊相连，和院墙一起，围成封闭式院落。

禅茶之院中央有景茶山流水，水景结合微地形的设计，大量运用宛转的曲线，山水从上往下一层层跌下，形象地塑造出"流水"的空间意境。

漪云亭坐落于湖石叠起的山峦之上，清泉于山间溢出。跋步其上，露白风清，茶香清冽，于此间陶醉沉浸，俯瞰深深庭院，赏全园之景。

uided by the residential philosophy of "Oriental tea charm, heaven in water and cloud", this project combines space change
ith Zen tea culture, which carries forward construction spirit of Jiangnan gardens and at the same time inherits local vein
nnotation, making people inspired by the philosophy of "Same taste of Zen and tea, same nature of humans and universe"
hen they travel around mountains and rivers.

ased on the superior geographical conditions, the design exhibits the natural poetic dwelling and the oriental cultural
eritage. The building of courtyard space and the spirit of traditional space place become the focus and highlight of the design.
he design starts from the most exquisite "façade" and uses "three doors, in five rooms" width" as the command. It adopts the
ing's mansion gate shape" with the highest level and the widest space in the ancient gate. The screen walls are arranged in
hinese character "Eight" lines, and the fifty-five meter mansion gate is magnificant. In ancient times, the gate of this scale
annot be used in the houses of non princes and nobles. It is a symbol of the master"s prominent position. The ancient ritual
stem is highlighted by the door.

oth sides of the screen walls can enhance the popularity of the courtyard and help to accumulate money. The gate adopts the
ne palace nail gate, which embodies the highest standard of royal etiquette.

he courtyard of Zen tea adopts the layout of "four water return to hall", surrounded by ambulatory. "Four water return to hall"
a unique layout of Jiangnan dwellings in Anhui Province. Its architectural style idea comes from the philosophy of the unity
heaven and man in Huizhou cultural heritage. The individual buildings of "four water return to hall" style house take the
aditional "room" as the basic unit, and the individual buildings are connected by galleries, together with the courtyard walls, to
rm a closed courtyard.

the center of the Zen tea garden, there is a scenic tea mountain and flowing water. Combining the design of waterscape
th micro topography, a large number of curves are used. The landscape falls from top to bottom, vividly creating the spatial
ood of "flowing water".

冷香亭 Lengxiang Pavilion

观山廊 Guanshan Gallery

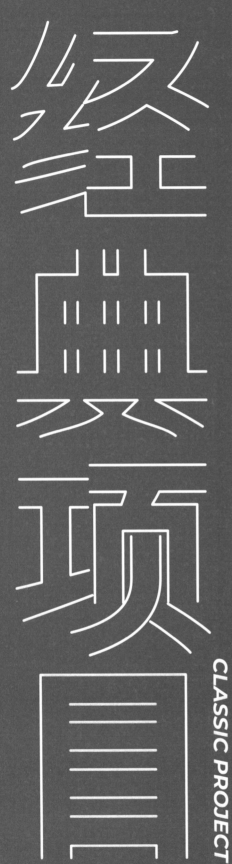

经典项目

CLASSIC PROJECT

住宅 – RESIDENTIAL

商办酒店 – COMMERCIAL - OFFICE - HOTEL

公共 – PUBLIC

住宅

RESIDENTIAL

怎样的载体才能更好地承载我们的生活，我们的设计能否为这份承载多做点什么？这一次，设计师尝试着在风格的拿捏、功能的平衡、理念的升级之外，让设计为"空间承载生活"多做点什么。

What kind of carriers can better carry our lives, and can our design do something more for the carrier? This time, the designer tried to make the design do something more for the philosopy of "space to carry life" apart from the style, function balance and concept upgrade.

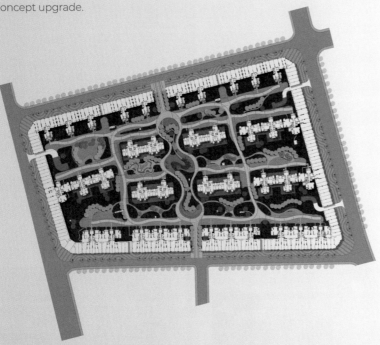

桂林融创栖霞府北苑
Guilin Sunac Qixiafu North Garden

设计时间：2018 年
项目委托：融创集团
项目地址：广西 桂林
项目规模：59400 ㎡

Design time: 2018
Entrusting party: Sunac Group
Project site: Guangxi, Guilin
Project scale: 59, 400 ㎡

▲ 亲影连廊 Shadow Gallery

峰环野立，一水抱城流。在了解了桂林山、人关系的发展历史以及社区现状后，设团队便以水的各种曲线形态和特征为灵感，十景观空间和节点，让人们更好地了解自之于生活的重要性，重塑人与自然的关系。设计语言上优雅地将自然元素与现代景观十编织在一起，实现自然元素与现代材料多式相互融合。

挂林的锦石奇峰为景观元素提取抽离，叠为峰，水景为江，将一幅山雨欲来、烟雾曼的中国山水画卷呈现出来。取河滩蜿蜒于、忽现于水之形，用叠级的景观手法加表现，营造出河岸浅滩、放歌垂钓的休闲意。

本景观设计以流动的曲线、国画的笔触，在场地内展开。水景流过整个场地，时而是迷人的梯田式水平轴景观，时而是优雅的水中庭院，最后汇集到中心水面。沿水边的树木将自然的浪漫气息融入人造的城市景观，呈现出独具趣味的地方文化。

入口处景观设计以桂林的龙胜梯田为景观元素进行提取抽离，将弧线、层级景观作为景观表现手法提出。增强入口的仪式感和参与感，突出休闲度假的氛围，打造出宽阔的水平轴景观视线。

曲径至此，顿时色彩明艳，该场地定位为活动区块，取意七星岩中天柱摘星，用明艳的色彩与适宜的活动设施配比，构建活力运动社区。

▲ 亲影连廊
Shadow Gallery

◀ 南入口景观区
South Entrance Landscape Area

▲ 亲影连廊 Shadow Gallery

housand of peaks stand around the wild, and one water flows around the city. After understanding the development history of Guilin mountain, water and human relationship as well as the current situation of the community, the design takes various curve forms and characteristics of water as inspiration to design landscape space and nodes, so that people can better understand the importance of nature to life and reshape the relationship between human and nature. In the design language, the natural elements and modern landscape design are weaved together gracefully, and the natural elements and modern materials and forms are integrated with each other. The overall landscape design is carried out in the field with a flowing curve and in Chinese painting style. The waterscape flows through the entire site, from a charming terraced horizontal axis landscape to an elegant underwater courtyard, and finally gathers into the central water surface. The trees along the waterside blend the natural romance into the artificial urban landscape, presenting a unique and interesting local culture.

Taking the beautiful rocks and grotesque peak of Guilin as the landscape element, the laid stone is the peak and the waterscape is the water, depicting a picture of wind-sweeping and mist-filled Chinese landscape painting. Mimicking the river beach twists and turns and the shape of water, and expressing with

▲ 住宅入户 Residential Home

an overlapping landscape technique, this project creates a leisurely artistic conception of singing and fishing along the shoal of river banks.

The landscape design at the entrance extracts the elements in Longsheng Terrace in Guilin, and the arc and level landscapes are selected as landscape technique of expressions. Thus, the sense of ritual and participation at the entrance is enhanced, the atmosphere of leisure and holiday is highlighted, and a wide horizontal axis landscape is

created.

Winding path ends here, the color suddenly becomes bright, and the venue is positioned as an active block, representing the implication of Picking up Stars on the Sky Column (a scenic spot) in the Seven Star Cave. In addition, the bright colors match the activity facilities, building a dynamic sports community.

华侨城香山美墅，位于深圳市南山区华侨城片区，东面为香山里等高档小区，东南面为成熟的波托菲诺社区，南面香山西街一路之隔为白石洲，西面为沙河东路及名商高尔夫球场，北面为北环大道，交通便利，地理位置优越。

华侨城始终以绿色生态为基点，以优质生活的创想家为理念，创造了诸多中国生态城市样本。旗下天麓、波托菲诺、纯水岸、天鹅堡、香山里等地产作品，融旅游、生态、文化于一体，成为繁华都市里诗意的栖居地。

Located in OCT, Nanshan District, Shenzhen, OCT-Chanson Villa is adjacent to Xiangshanli and other high-end residential areas in the east, well-developed Portofino community in the southeast, Baishizhou facing Xiangshan West Street in the south, and Shahe East Road and famous golf course in the west, and North Ring Road in the north, which gains access to convenient traffic and excellent geographical location.

Strictly based on green ecology, and following the concept of high-quality life, OCT has created numerous ecological city samples in China. Subordinated Tianlu, Portofino, Pure Water Bank， Swan Stone Castle, Xiangshanli and other real estate become a poetic habitat in the bustling city integrating tourism, ecology and culture.

深圳华侨城香山美墅
Shenzhen OCT - Chanson Villa

设计时间：2014 年
项目委托：华侨城集团
项目地址：广东 深圳
项目规模：58000 ㎡

Design time: 2014
Entrusting party: OCT Group
Project site: Shenzhen, Guangdong
Project scale: 58, 000 ㎡

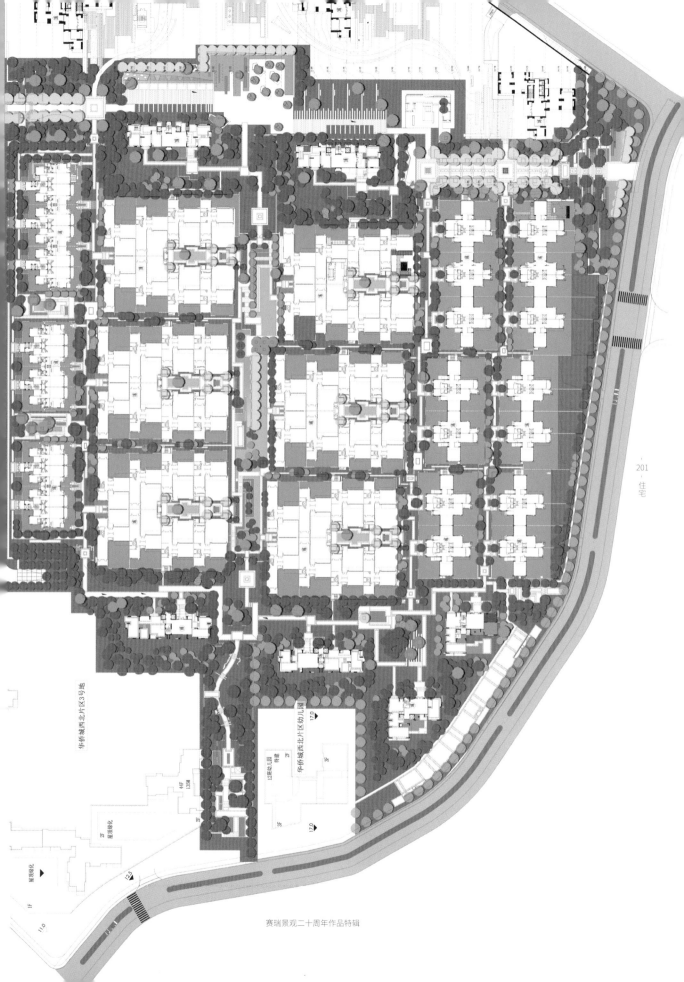

华侨城西北片区3号地

华侨城西北片区幼儿园

201 · 住宅

大隐城央，墅级院落，
不曾不在，经典传承。

Da Yin Cheng Yang, villa-level
courtyard,
Never been, classic heritage.

▲ 廊苑 Gallery

▲ 廊苑 Gallery

"落霞与孤鹜齐飞，秋水共长天一色"

"The falling clouds and the lonely are

flying together, and the autumn water

is always a day"

▲ 林下休闲场地 Underwood Recreation Ground

繁杂吵闹的都市里，在这里，给你一份心
的栖息地，左手繁华，右手宁静，有一种
活，叫华侨城。

the hustle and bustle of the city,
ere, give you a habitat of the soul. The
t hand is prosperous, the right hand
quiet. There is a kind of life, called
CT.

▲ 林荫道路 Tree-lined Road

项目采用传统造园原则营造园林，致敬经典，缩印留园；咫尺山林，小中见大，于方寸之间构建园林生活，以自然山水景观为基调，创造礼序与韵致空间。

整个空间布局采取一轴三进院的设计手法，入口布局中轴对称，序列严整。三进院落层层递进，院落、园林层层深入，尺度恢弘，等级分明，渲染高潮迭起的空间氛围。将传统的文化与园林要素进行剥离再组织，植入诗意典雅而不乏现代性的生活态度，再现一处别有气韵的国粹大宅。诠释东方美学的同时，营造古代达官贵人所生活的，高墙阔府、显赫门第的王府大宅，匠造一处方正典雅、围合而居的东方巨著。

The project adopts the principle of traditional gardening to create a classic garden, a miniature of Lingering Garden; the garden life is quintessential with the natural landscape as the keynote to present order and charm.

The whole space layout adopts the design method of one axis and three courtyards, with the axis being symmetrically designed against the entrance in an orderly array. The three courtyards are dynamic in structure, magnificent in scale, distinct in rank and climatic in space as a whole. The traditional culture and garden elements are properly reorganized into the poetic, elegant, and modernistic attitude of life, as a classic mansion with unique charm is reproduced. As an embodiment of the oriental aesthetics, it is worthy of comparison with an ancient palace for princes and nobles. And it is a Chinese style enclosed mansion featured by elegance and dignity.

宁德建发天行泱著

Tianxingyang Housing Project, Jianfa, Ningde

设计时间：2019 年
项目委托：建发地产
项目地址：福建 宁德
项目规模：5724 ㎡

Design time: 2019
Entrusting party: C&D Real Estate
Project site: Fujian, Ningde
Project scale: 5, 724 ㎡

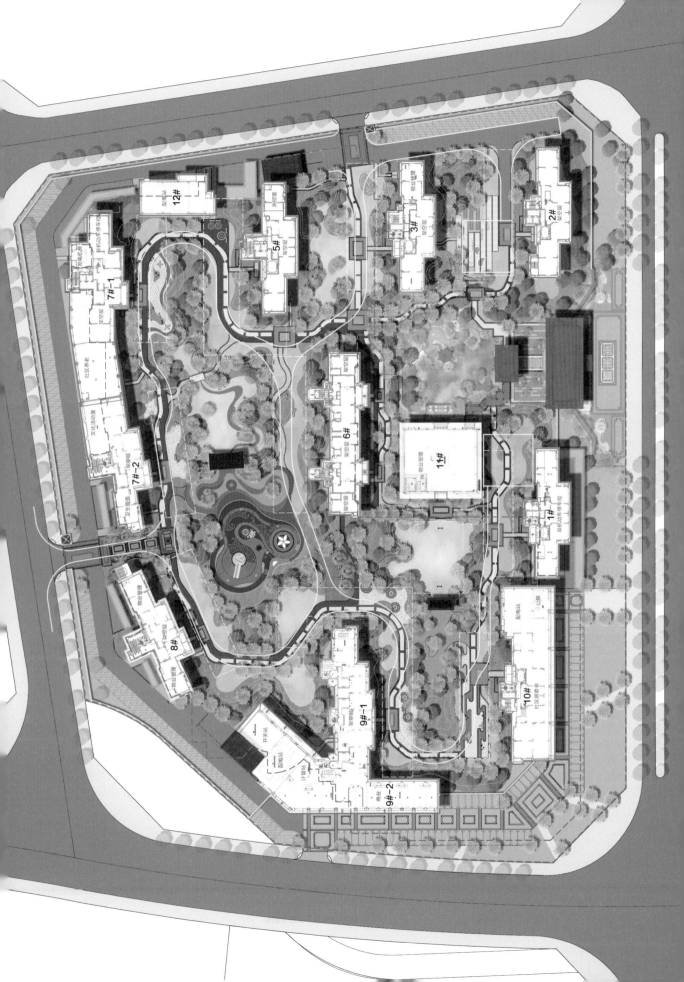

▲ 博雅阁 Boya Court

▲ 一进博雅阁 First Entrance: Boya Court

▲ 二进畅和园 Second Entrance: Changhe Garde

▶ 三进澄泓园 Third Entrance: Chenghong Garde

项目位于民乐大道以东，宝鼎路以北，鼎湖大道以东，创业路以北，项目地块路网完整，毗邻西江，有着发达的湖泊和河水系资源，是"粤港澳大湾区发展带"、"广佛经济圈"的重要组成部分。

本案设计以"玖致归一，盘龙之境"为主题理念，将情、景、境、雅、品、典、细、心、感等特性发挥到极致，融入整个项目之中，注重空间氛围与细节的控制，打造品牌社区。

The project is located at the east of Minle Avenue and the north of Baoding Road, the east of Dinghu Avenue and the north of Chuangye Road. The project plot has a complete road network and is adjacent to Xijiang River, thus boasting developed lake and river resources. It is an important component of "Development Belt of Guangdong-Hong Kong-Macao Greater Bay Area" and "Guangzhou-Foshan Economic Circle".

The design sticks to the idea of "pursuing perfection and attaining lofty realm" and integrates various characteristics into the overall project, including situation, emotion, elegance, taste and feeling. It emphasizes spatial atmosphere and detail control, and aims to create a brand community.

肇庆龙光·玖龙山
City Artery in Zhaoqing City

设计时间：2018 年
项目委托：龙光地产
项目地址：广东 肇庆
项目规模：7492 ㎡

Design time: 2018
Entrusting party: Logan Estate
Project site: Zhaoqing, Guangdong
Project scale: 7, 492 ㎡

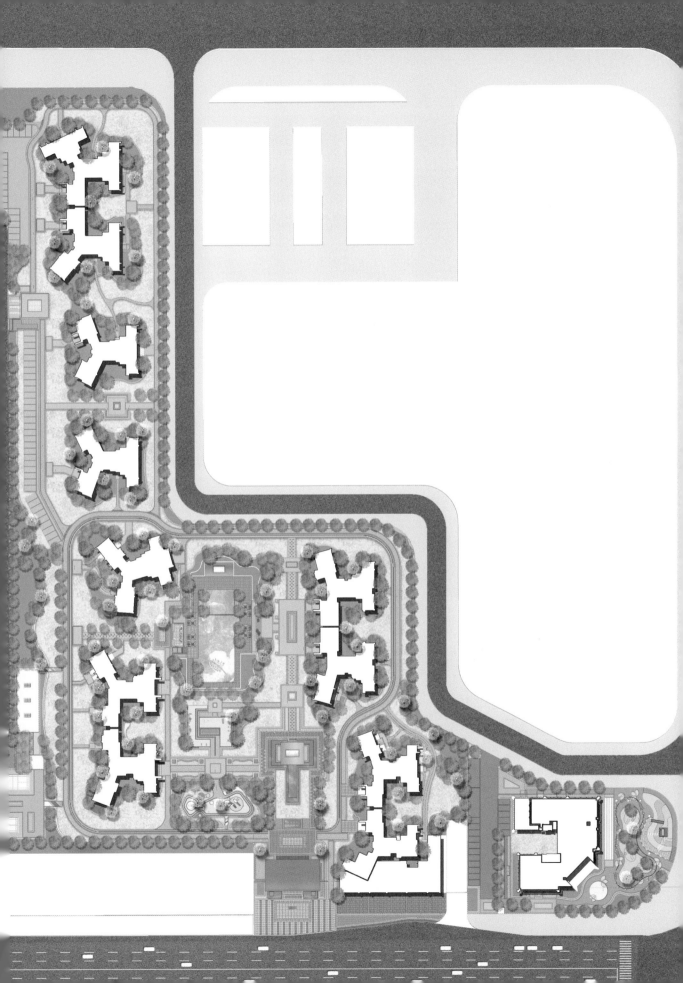

CSC 20th SELECTED PROJECT

项目着重打造"由繁尘入雅境、自城市归家园"的归途理念，售楼部与小区展示景观，通过"迎宾"、"初探"、"印象"、"展卷"、"入境"和"融境"六个部分的景观空间，层层递进，引导人们与空间的互动，营造一个有设计感、情景感和归属感的空间氛围。以尺度变化、绿化围合等方式营造积极互动的邻里关系，创造交流、路遇的邻里生活情景，打造生态自然、轻奢优雅的多彩共享、充满人情味的交流空间。

采用现代轴线对称手法进行设计，力求景观与简约现代的建筑风格达成一致性，使得建筑与景观相融合。建筑设计强调竖向线条的立体感，色彩搭配轻快明朗，富有时代气息。

The project focuses on creating a return concept of "from dust to environment and from city to home". The sales department displays the landscape through the six parts of the landscape space of "welcome", "preliminary exploration", "impression", "exhibition", "entry" and "fusion". The project gradually leads people to interact with the space, and creates an atmosphere with a sense of design, scene and belonging. By means of scale change, green enclosure and other ways to create a positive and interactive neighborhood relationship, a neighborhood life scene of communication and encounter, and a colorful sharing space of ecological nature, full of human emotions.

It is designed by combining the modern techniques and axis symmetry, so as to make the landscape in line with the concise and modern architectural style and integrate the architecture with the landscape. The architectural design emphasizes the three-dimensional sense of vertical lines and has bright and brisk color allocations, which is filled with the flavor of the times.

入口门楼
Entrance Gate

▶ 叠级水景
Drop in Waterscape

济南万达城的景观风格与建筑风格融合共生，设计意境的再造升华，是地域文化与现代风格的碰撞、中式情怀与都市繁华的融合。本项目将打造兼具"文化性、标识性、生态性、功能性、安全性"的高品质休闲度假城。

The landscape style and architectural style of Jinan Wanda City are well matched, and re-construction and upgrade of artistic conception of the design fully represent the collision of regional culture and modern style, and the combination of Chinese culture and urban prosperity. This project aims to create a high-quality leisure resort city integrated with cultural property, identification, ecology, functionality and safety.

济南万达城
Jinan Wanda City

设计时间: 2016 年
项目委托: 万达集团
项目地址: 山东 济南
项目规模: 20000 ㎡

Design time: 2016
Entrusting party: Wanda Group
Project site: Jinan, Shandong
Project scale: 20, 000 ㎡

项目位于山东省济南市历城区，与山东建
大学、济南国际会展中心、济南奥体中心
邻；交通便利，北邻 G309 国道荣兰线；
济南东火车站、济南遥墙国际机场形成半
时生活交通圈；周边韩仓河东城景观带，
观资源丰富，发展潜力巨大。
"玉盘凝珠，碧水生莲"从荷塘的荷叶、
珠、莲花、湖水中提取元素，寻找它们之间
出淤泥而不染的精神内涵，创造它们真实的
触感，并延伸到人与自然和谐相融。在现
设计手法演绎下，从而给人一个从传统到
代的回归展望，让人久久回味。

▼ 风雨连廊 Storm Corridor

▲ 展示区广场
Exhibition Area Square

赛瑞景观二十周年作品特辑

Located in Licheng District, Jinan City, Shandong, the project is adjacent to Shandong Jianzhu University, Jinan International Conference Center and Jinan Olympic Sports Center. Closed to G309 national highway - Ronglan Road in the north, convenient in traffic, it only takes half an hour to arrive at Jinandong Railway Station and Jinan Yaoqiang International Airport. Surrounded by East Dongcheng Landscape Belt of Hancang River, it is rich in landscape resources with huge development potential.

"Water drops solidify into beads in the jade plate, lotus grows in the green water",the design extracts elements from lotus leaves, lotus beads, lotus flowers and lake water in the lotus pond, seeks for the spiritual connotation of "the unstained emerging from the filth" between them, creates the true touch and extends it to people and nature in harmonious coexistence. Based on modern design methods, people are allowed to return to the past, making people remember it for a quite long time.

风雨连廊 Storm Corridor

海马公园二期位于郑州市东风路与商都路交汇处东北，作为 CBD 和中央政务区的功能支撑区，有 CBD 的后花园之称，建有香花森林、四季花林、彩叶森林、生态森林四大森林组团，大面积的景观绿化，堪比公园，公园生活，由此开启。

Located at the northeast to the crossing of Dongfeng Road and Shangdu Road, Zhengzhou City, Haima Park (Phase II) is known as the backyard garden of the CBD serving as the functional support area of CBD and central government affairs district. It consists of Fragrant Flower Forest, Four Season Flower Forest, Colorful Leaves Forest and Ecological Forest. In the large area of landscape planting, the life in the park begins.

郑州海马公园二期
Zhengzhou Haima Park (Phase II)

设计时间：2012 年
项目委托：海马集团
项目地址：河南 郑州
项目规模：53000 ㎡

Design time: 2012
Entrusting party: Haima Group
Project site: Zhengzhou, Henan
Project scale: 53, 000 ㎡

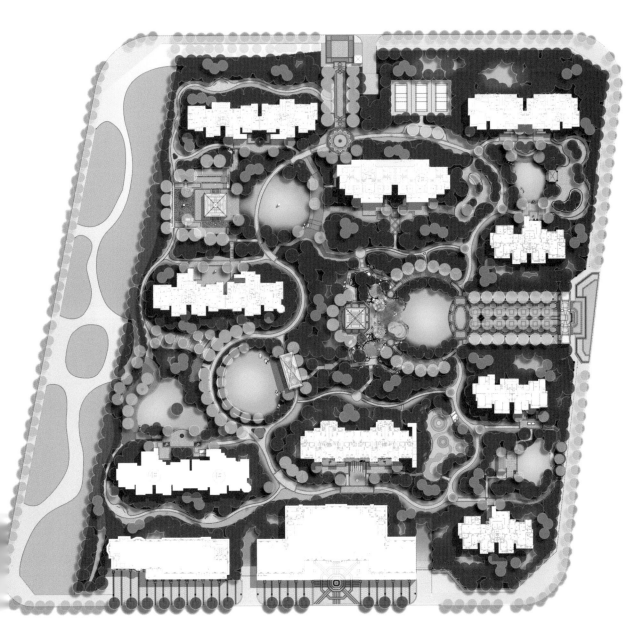

▲ 住宅入口 Residential Entrance

▲ 特色水景 Characteristic Water Feature

CSC 20th SELECTED PROJECT

马公园景观揉和了现代古典主义风格及简元素于景观设计中，结合 ARTDECO 建筑格，运用自然造景为主，强调景观的功能，表现森林、坡地的视觉效果，体现人性、然、简洁、实用的景观设计思想，使生活有古典色彩，亦可与大自然融合，让人可空间中寻回古典的、优雅的、宁静的气息。"建筑放在公园里"是景观设计的主要理念。"公园"为主题，建立起鲜明的社区形象，过"中央公园"、"邻里公园"、"亲子园"、"健身公园"，创造强有力的场所和自然归属感。

The landscape in Haima Park adopts modern classicism and austere European style in the landscape design. Its design, combined with ARTDECO style and dominated by natural landscaping, attaches importance to the functionality of the landscape to demonstrate the visual effect of forest and sloping fields. The landscape design idea, reflecting humanity, nature, simplicity and utility, enables the life to be classical and natural. People can regain the lost elegance and tranquility in traditional lives. "Putting structures in the park" is the main design principle in the landscape design. With "park" as the theme, a distinct community image is established, and the strong sense of space and natural belonging are created through different themes including "Central Park", "Neighborhood Park", "Parent-child Park", and "Exercise Park".

▲ 景观亭 View Pavilion

酒 商
店 办

COMMERCIAL - OFFICE - HOTEL

会展湾·南岸位于深圳市重点规划片区会展新城，总建面约 300 万平方米的会展主题综合体。项目设计注重城市空间及环境的互相关联，强调空间的连续组织及关系，塑造人们对环境空间的视觉感受。通过对光、轴线、竹子、砂石、水幕等元素的提取有机地结合在一起，构建以人的景观感知为中心的体验空间序列。在人们创造空间美的同时，空间美也在改变着人们的生活品质。在深圳这座国际化城市，会展湾·南岸宛如一颗璀璨耀眼的明珠屹立于粤港澳大湾区之心。

The Expo Bay·South Coast is located in Shenzhen key planning district, Expo New City. As the initial project of the Expo themed complex covering an area of 3 million square meters. It is not only an exhibition center but also an art space. More efforts are made in the interrelation of city space and environment in the project design to emphasize continuous structure and relation of space and shape people's visual sense on environment space. Elements like light, axes, bamboo, sand rock and water curtain are extracted and blended together to build the experimental spatial sequence centered on people's landscape perception. When people build the spatial beauty, the latter is at the same time improving people's life quality. In the international city Shenzhen, the Expo Bay · South Coast stands in the center of Guangdong-Hong Kong-Macao Greater Bay Area like a dazzling pearl.

深圳会展湾·南岸

Shenzhen Expo Bay · South Coast

设计时间：2018 年	Design time: 2018
项目委托：深圳市招华会展实业有限公司	Entrusting party: Shenzhen Zhaohua Exhibition Industry Co., Ltd.
项目地址：广东 深圳	Project site: Shenzhen, Guangdong
项目规模：37000 ㎡	Project scale: 37, 000 ㎡

针对不同的城市界面，我们在设计中提出了不同的设计策略。北侧部分布置公寓与酒店，在满足公寓及酒店景观最大化的同时，打造项目连续的整体的大尺度形象，并且为沿河的城市景观提供了现代化的建筑界面。景观设计希望把建筑、环境和社会结合在一起，当作一个有机整体去设计。

By targeting at different urban interfaces, we propose different design strategies in our design. In the north side, we arrange the apartment and hotel. While realizing the maximum of apartment and hotel landscape, the design creates the continuous and overall large-scale image of the project, and provides the architectural interface for the urban landscape along the river. Landscape design hopes to combine architecture, environment and the society and regard them as an organic whole. Consciously organize the overall order, make the parts perceived by people in an orderly way, and highlight the beauty of coordination in changes.

◀ 酒店入口跌水 Hotel Entrance Falls

酒店下沉广场
Hotel Sinking Square

▲ 特色休息座椅 Featured Rest Seat

生态步道 Ecological Trail

该项目是典型的商业景观设计项目，其中心
内容均是为商业场所的运作而服务，强调人
流与商品或有偿的服务项目的交流与互动。
在这一前提主导下，结合景观设计的美学原
理和布局方式，来满足丰富业态的需求。

It is the typical commercial landscape
design project, and mainly serves for
the operation of commercial. It
emphases interaction of paid service
for people and commodity. Dominated
by the premise and combining the
aesthetic principles with layout pattern
of landscape design, it meets the goal
of enriching industry forms.

郑州建正东方中心
Zhengzhou Jianzheng Oriental Center

设计时间：2011 年
项目委托：河南建正房地产有限公司
项目地址：河南 郑州
项目规模：29304 ㎡

Design time: 2011
Entrusting party: Henan Jianzheng Real Estate Co., Ltd.
Project site: Zhengzhou, Henan
Project scale: 29, 304 ㎡

景观设计将构思的基本理念结合于建筑之中，四栋塔楼错落相对，互相分裂又互相联系，形成自然、和谐、统一的关系。方形塔楼通过网络的组建构成立体网络的机理效果；景观设计基于项目文化背景及建筑现状，展现独一无二的构思，以"光耀中原"和"辐射东方"为设计思想，将铺装以"东方中心"向外辐射，强调主要入口广场的同时，关联四栋塔楼，形象彰显东方中心商业氛围的活力与开放，打造出本土化、具有文化属性并能唤起市民归属感的休闲活动场所。

ndscape design combines the basic idea of the concept into the building. The four towers are scattered and opposite, parated but connected with each other, forming a natural, harmonious and unified relationship. The square tower forms the echanism effect of three-dimensional network through the establishment of network; based on the cultural background d architectural status of the project, the landscape design shows a unique idea, taking "shining in the Central Plains" and diating to the East" as the design idea, the pavement will radiate outwards from "Oriental Center", while the main entrance uare will be highlighted, four towers will be associated, and the image will show the commercial atmosphere of the Oriental nter. It is vital and open to create a local leisure activity place with cultural attributes and evoking the sense of belonging nong the citizens.

▲ 景观通廊 Landscape Corridor

◀ 斜面种植池 Bevel Planting Pond

▶ 中心休息广场 Central Rest Square

CSC 20th SELECTED PROJECT

营造可持续的绿色生态办公区，打造引领时尚、彰显科技的智慧型商业区；延续客家文化内涵，造就人文典范标杆；倡导现代简约、高端品质的生活方式，是我们的设计目标。坪山区位于广东省深圳市东北部，总面积为 168 平方公里，是深圳市东部主要工业基地。坪山区东靠惠州大亚湾石化城，南连具有优美原生态的大鹏半岛，西邻世界最大的单体港——盐田港，北面是商贸发达、配套齐全的龙岗中心城，是深化深莞惠合作的重要战略节点。发展潜力巨大，可以为深圳未来的产业发展特别是高科技产业发展提供战略支撑。

"Create the sustainable office area of green ecology and create the smart business district that leads fashion and highlights technology; carry out the culture of the Hakkas and build the cultural model; advocate the modern and concise lifestyle with high-end quality" are our design objectives. This project is located at Pingshan District, the northeast of Shenzhen, Guangdong Province, with the total area of 168 square kilometers. It is the main industrial base in the east of Shenzhen. Pingshan District is close to Petrochemical City, Daya Bay, Huizhou in the east, Dapeng Peninsula with the beautiful original ecology in the south, the biggest monomer port Yantian Port in the west, and Longgang Central City with developed commerce and complete supporting facilities in the north. It is an important strategic mode of deepening the cooperation among Shenzhen, Dongguan and Huizhou. With a huge development potential, it willprovide strategic support for the high-tech development, especially the future industry development of Shenzhen.

益田·益科大厦
Yitian · Yike Tower

设计时间：2017 年	Design time: 2017
项目委托：益田集团	Entrusting party: Yitian Group
项目地址：广东 深圳	Project site: Shenzhen, Guangdong
项目规模：15334 ㎡	Project scale: 15, 334 ㎡

▲ 商业入口广场 Commercial Entrance Pla

▲ 特色广场 Featured Squ

▲ 屋顶花园 Roof Garden

计灵感来源于科技、文化以及建筑三者，提了各自的结构布局特点，融合成整体性的观形式。以科技芯片的元素、客家文化凉的编织纹理，以及建筑立面的竖向特色空，呈现出科技文化的表达，以一个个舞动单元格象征前行的脚印一般，象征着科技速发展加快人类前行的步伐。将文化记忆织再现，有机组合融合为一个有序的整体，显出当下社会一种科技、艺术、文化相互合发展的趋势。

The design inspiration comes from technology, culture and architecture. The designer extracts the structural layout characteristics and integrates them into the holistic landscape form. It expresses the technological culture with the elements of technological chips, the summer hat weaving texture of Hakka culture and the verticalspace of the building. The dancing cells embody the forwarding footprints, which symbolizes that the rapid technological development speeds up humankind's paces. It weaves the cultural memories and organically integrates them into an orderly entirety, which shows the tendency of the integrated development of technology, art and culture in the current society.

本项目位于深圳沙嘴，临近红树林、深圳湾，拥有一线滨海景观。项目营造了热情、神秘、浪漫、怡人的特色滨海文化景观，在刚柔韵致的空间中感悟海洋文化气息。寻觅和回归人们心中自由体验生命的悦动和浪漫，感受物之尽藏，与大海晤谈。

The project is located at Shazui, Shenzhen, close to Mangrove and Shenzhen Bay, thus boasting the closest coastal landscape. It creates the enthusiastic, mysterious, romantic and pleasant special coastal cultural landscape, where people may feel the marine culture in the charming space. It seeks for and creates the pleasure and romance of freely experiencing life for people, so that they may feel the vast world and talk with the sea.

深圳绿景红树华府
Shenzhen Lvgem Hongshu Huafu

设计时间：2013 年
项目委托：绿景集团
项目地址：广东 深圳
项目规模：35000 ㎡

Design time: 2013
Entrusting party: Lvgem Goup
Project site: Shenzhen, Guangdong
Project scale: 35, 000 ㎡

◀ 商业分区
Commercial Divis

走出茂密的丛林，进入豁然开朗的领地。宽
敞开阔的廊架、细沙、棚林浪，体现空间的
延展与活跃，带给人们强烈的视觉冲击与独
特的吸引。形如海浪般起伏的界面和由海岛
形成自由组合的竖向空间，形成了富有动态
的景观特色。目的是创造人们可参与的空间，
打造一个富有个性、充满活力的多元化综合
生活场地。

Walk out of the leafy forest and enter an enlightening territory. The spacious and
wide corridor frame, fine sediment and waves embody the extended and active
space, and bring the strong visual shock and unique attraction to people. The wave-
shaped interfaces and the vertical space freely formed by the island constitute the
dynamic landscape characteristics. It aims to create a space where people may
participate in, and create an individualized, vigorous, diversified and integrated
living site.

屋顶花园阅海堂
Roof Ocean Garden Yuehaitang

屋顶泳池
Roof Pool

屋顶花园映月水台
Roof Moon Water Courtyard

深圳龙华创想大厦项目总体定位为深圳新中心区地标综合体，是以商业办公为主要功能，同时兼有休闲、娱乐等功能的复合型地产项目。复合房地产将房地产开发与创造消费者生活方式密切结合，为消费者创造出充分体现生活感受和文化价值的复合人居生活。

The overall positioning of Shenzhen Longhua Creativity Building is the landmark complex of Shenzhen new central area. It is a complex real estate project with commercial office as its main function and combined with leisure, entertainment and other functions. Compound real estate combines real estate development and consumer lifestyle creation to create a compound living for consumers that fully reflects their life feelings and cultural values.

深圳龙华创想大厦
Shenzhen Longhua Creativity Building

设计时间：2017 年
项目委托：深圳华侨城置业投资有限公司
项目地址：广东 深圳
项目规模：24926 ㎡

Design time: 2017
Entrusting party: Shenzhen OCT Real Estate Investment Co., Ltd.
Project site: Shenzhen, Guangdong
Project scale: 24, 926 ㎡

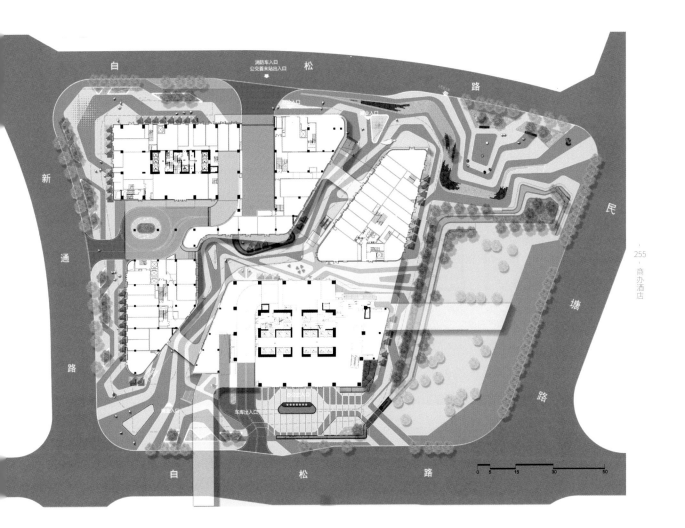

白 松 路

白 松 路

新 通 路

民 塘 路

消防车入口
公交首末站出入口

车库出入口

0 5   15   30   50

▲ 展示区入口
　 Exhibition Area Entrance

◀ 光影广场
　 Light Shadow Square

CSC 20th SELECTED PROJECT

观设计灵感来源于能够不断重生的自然力，将溪水、山石、植物、花镜、岩洞、山的阳光、瀑布等自然环境元素巧妙地融入观设计之中，"溪谷"的设计带来了众多体验要素。

山、谷、云、林四个自然元素作为主要元构成，打造"映水藏山·溪谷探秘"的生景观体验。这种体验手法的空间再造，为目展现了全新的生态景观。山：绿岛广场，有洞天；谷：银河瀑布，峡谷探秘；云：泳池，镜像云谷；林：林间星空，绿树荫。

The inspiration of landscape design comes from the natural forces that can be constantly reborn. The natural environment elements such as streams, rocks, plants, flower borders, caves, mountain sun, waterfalls, etc. They are skillfully integrated into the landscape design, and the design of the valley brings numerous experience elements.

With the four natural elements of mountain, valley, cloud and forest as the main elements, the ecological landscape experience of "reflecting water and concealing mountains· exploring secrets in the valley" is created. The spatial reconstruction of this kind of experience technique reveals a new ecological landscape for the project. Mountain: The green island square, here is an altogether different world; Valley: Galaxy waterfall, here is valley-exploring. Cloud: Oxygen bar and pool, mirror image and cloud valley; Forest: Starry sky in woodland, and trees shade the street.

广场中布置火烈鸟主题雕塑作为亮点，展示两种主题特质的火烈鸟，打造商业卡通形象标签。主题一：优雅、尊贵，休憩的火烈鸟，或静立水中，或低头饮水，身姿绰约、优雅尊贵。主题二：飞翔、共舞，跳舞表达集体的愉悦，腾空展示个体的才能，所到之处必将是人流聚焦之地，吸引无数目光。设置五种飞翔姿态，展示火烈鸟从起飞到高飞、从滑翔到群飞的过程，体现扬梦起飞、共享未来的商业精神气质。基于项目的总体定位，延续建筑"泛公园设计，都市微森林"的概念，意在采用现代风格的景观设计来打造一个生态性商业购物公园，使其成为深圳新中心区地标综合体。

The square is decorated with the flamingo themed sculpture as the highlight, displaying two characteristics of flamingo, to create the commercial cartoon image label. Theme one: Elegant, noble, resting flamingo, either standing still in the water, or bowing to drink water with a graceful posture, elegant and noble. Theme two: Flying, dancing togetherto express the collective joy, showing the individual talent in the sky, everywhere they go will be the focus of the crowd, attracting countless eye balls. Five flying positions to show flamingo from takeoff to flying high, from gliding to flying as a flock, which reflect the business culture of flying with dreams and sharing the future. Based on the overall positioning of the project, the concept of "general park design and urban micro-forest" of the building is carried forward. It intends to create an ecological and commercial shopping park with modern landscape design, making it the landmark complex of Shenzhen new central area.

▶ 商业区中心
Commericial Center

▶ 裙房塔楼
Skirt House

东莞天安数码城坐落于东莞市南城区白马地块，东莞运河东南岸。地块的西南面接广深高速公路，交通便利，地理位置优越。东莞天安数码城拟建成 21 世纪的城市信息科技园，主要功能包括以数码信息产品的研发和展销为主的创业和技术交流中心，以及相应公共和必要居住配套设施等。

Dongguan Tian'an Cyber-City is located in the Baima block of Nancheng District, Dongguan City, on the southeastern shore of the Dongguan Canal. The southwest side of the plot is connected to the Guangzhou-Shenzhen expressway, with convenient transportation and superior geographical position. Dongguan Tianan Digital City plans to build a 21st century urban information technology park. The main functions include the establishment of a digital information product research and development and technology exchange center, and the corresponding public and necessary residential facilities.

东莞天安数码城
Dongguan Tian'an Cyber-City

设计时间：2012 年
项目委托：天安集团
项目地址：广东 东莞
项目规模：30771 ㎡

Design time: 2012
Entrusting party: Tian'an Group
Project site: Dongguan, Guangdong
Project scale: 30, 771 ㎡

Green Island Atrium District

通过景观与建筑设计的融合与呼应，形成别具特色、如城市绿洲般的数码社区环境。以东莞的人文地理为文脉，打造精致的现代景观主题和特色空间，同时展现未来高科技趋势。提供一个舒适并留下美好记忆的户外空间，创造强有力的场所感和社区归属感。

建立合理、层次分明的公共半私密共享交流空间，用于一系列的户外休闲与商用机遇，同时也与城市开放空间系统衔接。创造本期开发的景观特色，同时采用一些统一的景观元素、铺地材料和植栽品种等，使本区域景观设计既能与整个科技园的风格一致，又具有其独具一格的特色。不论白天或夜晚，为园区工作人员和来访者营造以人为本、可持续发展的良好户外环境。

Through the integration and echo of landscape and architectural design, a digital community environment with unique characteristics such as urban oasis is formed. Taking Dongguan's human geography as a context, we will create a refined modern landscape theme and features, while reflecting the future of technology trends. Provide an outdoor space that is comfortable and leaves a good memory, creating a strong sense of place and community.

Establishing a reasonable, hierarchical public, semi-private shared communication space for a range of outdoor leisure and commercial opportunities, as well as the urban open space system lies in the center of the design concept. Creating the landscape features developed in this period, and adopting some unified landscape elements, paving materials and planting varieties, the landscape design of the region can be consistent with the style of the entire science park, and has its unique characteristics. On the day or night, we will create a good outdoor environment for people and visitors in the park to achieve people-oriented and sustainable development.

▼ 绿岛中庭区 Green Island Atrium District

水景 Water Feature

绿岛中庭区 Green Island Atrium District

儋州恒大海花岛
Danzhou Ocean Flower Island

设计时间：2015 年
项目委托：恒大集团
项目地址：海南 儋州
项目规模：423000 ㎡

Design time: 2015
Entrusting party: Evergrande Group
Project site: Danzhou, Hainan
Project scale: 423, 000 ㎡

花岛地处海南，属热带湿润季风气候，夏无
暑，冬无严寒，阳光充足，雨量充沛，气
宜人，自然条件优越。海花岛国际旅游度
岛建设成为以休闲旅游为特色的，集海洋
动、婚礼蜜月、商务会议、休闲度假于一
的国际级旅游区。
地域特征、气候特征、文化特征考虑，从
找到与景观相似或相近的元素，落实到方
设计当中，让游客从步入园区，即能融入
景氛围中，游客从建筑景观中优美的曲线
会到大海的柔情，从植物配置中尽染热带
林的热辣风情，从大型民俗表演中烙印下
史文化的记忆。整体景观为场地而生，是
界上独一无二的、创意新颖的、不可超越
。

The Ocean Flower Island is located
in Hainan and experiences a humid
tropical monsoon climate with super
hot summers and extremely cold
winters. This place is well endowed with
sufficient sunshine and precipitation.
The weather is pleasant and the natural
conditions are superior. The Ocean
Flower Island International Resort will
be constructed into an international
tourism area featured with leisure
tourism, and integrated marine
activities, wedding and honeymoons,
business conference and leisure
vacation.

Considering the regional features,
climatic characteristics and cultural
characteristics, we adopt such elements
into the landscape design, allowing
the tourists to be inside the scene once
they step into the garden, the tourists
will appreciate the tenderness of the
ocean from the beautiful curves of the
building,soaked into the hot customs of
the tropical rainforest and experiencing
the memory of historical culture
passed down from large-scale folk
performances. The entire landscape was
born for the site, it is a unique, novel and
unparalleled space in the world.

香水湾金缔度假酒店位于中国最大的热带海岛——海南岛。以其独有的延绵六十公里的热带海岸线、广阔的私家海滩和金海园林而得天独厚。该项目已多次被行业杂志收录，被列为经典的酒店景观设计范例。

该项目充分利用地方特色的素材，恰到好处地反映周边自然环境和地方文化元素，达到了设计的初衷。自然而不着痕迹的服务理念，足以让它成为度假酒店之设计经典。"最少的改变"和"恰当的梳理"才是最好的设计。

The Xiangshui Bay Golden Holiday Resort is located in Hainan Island, China's largest tropical island. With its unique tropical coastline of 60km, vast private beach and Jinhai garden, it is unique. The project has been included in industry magazines for many times and listed as a classic example of hotel landscape design.

By making full use of the materials with local characteristics and reflecting the surrounding natural environment and local cultural elements properly, the original intention of the design is achieved. The natural and traceless service concept makes this project a classic resort design. "Less is more"and "appropriate arrange" are the best design.

陵水红磡香水湾金缔度假酒店
Lingshui Hung Hom Xiangshui Bay Golden Holiday Resort

设计时间：2010 年
项目委托：红磡集团
项目地址：海南 陵水
项目规模：215000 ㎡

Design time: 2010
Entrusting party: Hung Hom Group
Project site: Hainan, Lingshui
Project scale: 215, 000 ㎡

▲ 休闲小道 Leisure P

▲ 别墅区鸟瞰 Aerial View of Villa Area

▲ 中心庭院 Central Courtyard　　▲ 热带风情泳池 Tropical Swimming Pool

店的景观设计以浓郁的东南亚风情为主，浪漫温馨的海岸生活气息融入景观设计中，造出难得的闲适的度假风情。

今多元化的生活让我们既不舍都市生活的速与便捷，又向往回归质朴的大自然，于城市中的度假酒店便成了联系城市与自然桥梁，它体现了这座城市的特性。我们为个酒店设计的景观，希望让走进这里的人受到故事性和归属性的空间，它需要像家样温馨和随性，让来这里度假的人们能觉自己就属于这里。

The landscape design of the hotel is in rich Southeast Asia style, which integrates the romantic and harmonious coastal life into the landscape design, creating a valuable and leisure vacation destination.

Today's diversified life makes us not only willing to give up the speed and convenience of urban life, but also yearn to return to the simple nature, so the resort hotel in the city has become a bridge connecting the city and nature. It embodies the characteristics of the city. The landscape designed for this hotel is to let people who comes here feel the story and the sense of belonging so it needs to be warm and casual like home.

酒店的规划理念是通过合理的分区，结合本项目自身的资源条件及主题定位，设定合理的功能分区来打造高档、精致、休闲的欧洲慢生活氛围。面对"快与慢"、"新与旧"的结合以及景观的主题定位，打造一个高品质的会所是最本质的诉求。

The planning concept of the hotel is to create a high-grade, exquisite and casual European slow living atmosphere through a reasonable division, combined with this project's own resource condition and theme positioning. Facing the combination of "fast and slow" and "new and old" and the theme of landscape, creating a high-quality clubhouse is the most essential appeal.

昆明华侨城高尔夫酒店
Kunming OCT Golf Hotel

设计时间：2011 年
项目委托：华侨城集团
项目地址：云南 昆明
项目规模：58800 ㎡

Design time: 2011
Entrusting party: OCT Group
Project site: Kunming, Yunnan
Project scale: 58,800 ㎡

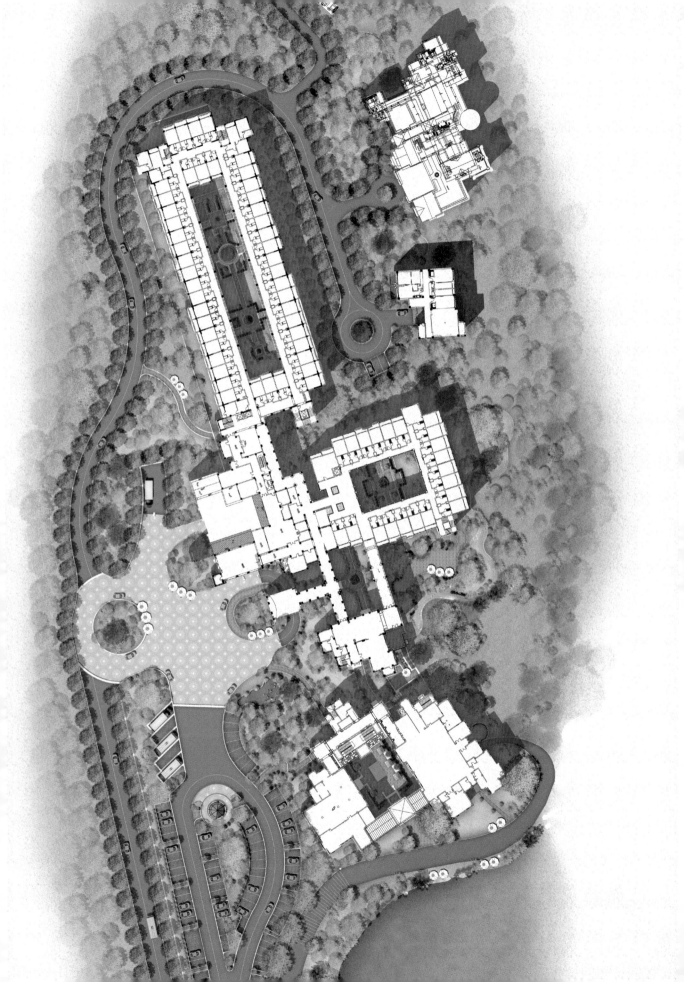

整体的景观作风延续南加州风格，质朴
自然的外观、粗犷的手工艺。除了外观
质感、温暖的外墙色彩，再加上宽敞的
庭院和景观露台，整体设计清新、明快、
有层次的叠落露台，供给观景平台，使
建筑内外的景观相互融会。选择本土多
种植物多层次搭配，形成亚热带休闲式
滨水性植物景观，营造一种自由、休闲、
自然的南加州风情生活。

结合本项目的自身条件及地理地势，打
造出自然、生态、低调而高贵的南加州风
格的高尔夫酒店，在缓坡、浅丘、清流
的地貌里，简约、明快的低密度绿化群
与现代欧风的商业区相辅相成，营造出
典型的加州田园特点。在这里度假、旅
游既是生活的态度，更是身份的象征。

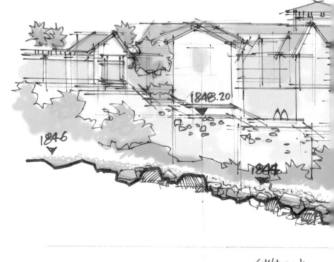

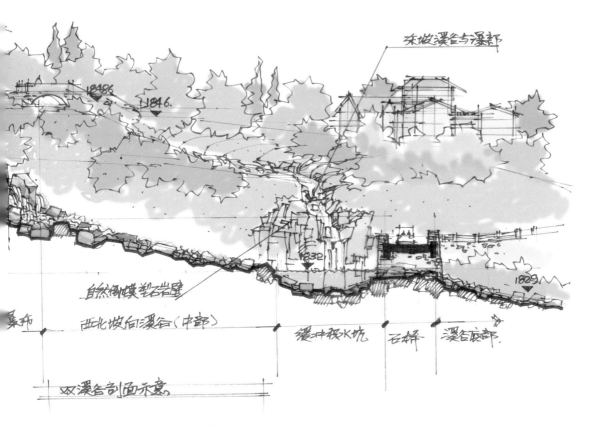

自然风貌塑石岩壁

西北坡白溪谷（中部）

瀑布

缓冲积水坑

石桥

溪谷底部

双溪谷剖面示意

朱坡溪谷与瀑部

▲ 庭院景观
Courtyard View

▼ 入口景观
Entrance Landscape

The overall landscape style continues the Southern California style, with a rustic and natural look and rough craftsmanship. In addition to the appearance of the texture, the warm exterior wall color, plus the spacious courtyard and landscape terrace, the project is designed with a fresh, bright and layered terrace, which provides a viewing platform to integrate the landscape inside and outside of the building. Choosing a multi-layered combination of native plants to form a subtropical recreational waterfront plant landscape to create a life with free, casual and natural Southern California style.

Combining this project's own conditions and geographical terrain, it creates a golf hotel with natural, ecological, low-key and noble Southern California-style. Design adopts gentle slopes, shallow hills clear streams, and the simple, bright and low-density greening clusters to complement modern European commercial area, creating a typical California pastoral character. Traveling here shows your attitude of life, and is also a symbol of identity.

游览观光路线大道
Sightseeing Route

PUBLIC

长沙湘江大道风光带
Xiang River Road, Changsha

设计时间：2008 年
项目委托：长沙市政府
项目地址：湖南 长沙
项目规模：138000 ㎡

Design time: 2008
Entrusting party: Municipal Government of Changsha
Project site: Changsha, Hunan
Project scale: 138, 000 ㎡

项目以铁路为主题贯穿四个节点和一个线性
亲水平台的平面和空间设计。

观景平台：以红色与黑色工字钢构架与局部
玻璃顶的形式恢复火车南站的原有空间形态
肌理，以钢架与刻有火车南站历史内容的黑
色枕木形成观景广场中的景观雕塑群，以多
股铁路交叉的形式构成多个线型的悬挑观景
平台，以路基石为主要填充材料的铺地，以
亲水木制平台与悬挑观景平台一起构成具有
动感的四维观景空间。

八道码头：利用现有的台阶和遗迹中部遗留
建筑的结构部分，以黑色工字钢构架、黑色枕
木与透明玻璃构成观景平台和立体空间，兼
顾保护、安全、观景与沿江景观的多方面需
要，尽可能保留遗迹、弱化新设计要素。

火车主题广场和液化气码头，以"时尚根据
地"的形式、与临近的液化气码头一起延续火
车主题广场的铁路遗迹文化、更新广场的商
业内涵，共同构成这个节点的景观系统。广
场上的运动设施、火车茶座、工厂酒吧、水

中表演舞台和路基石地面铺装等景观元素都
使遗迹景观在唤起都市人文化记忆的同时焕
发出火一般的青春活力。

在景观结构上强调四点、一线、多层，重点
处理、线性连接、空间层次。再生火车南站
旧址的框架场景，重现车轨与船运码头；尊
重八道码头现状、架空栈道、玻璃的应用，
低调处理场地；重构火车广场与液化气码头，
植入时尚活动与极限运动，拓宽地下通道将
广场、河滩、码头连为一体，打造区域内的
时尚根据地；回归自然绿化为主的桥头绿地
与南郊公园融成一体；自由线性的亲水平台
与滨江背景林带连接了被新建干道割裂的山
与水；多层的空间景观效果来源于丰富、多
样和立体的景观结构。以模糊边界、肌理构
成和多层次过渡的手法弥补道路设计中遗留
的设计缺陷，使自然、道路和景观构成一个
和谐的系统。

The project is a railway-themed plane and spatial design that connecting four nodes and one linear water platform. The viewing platform restores the original spatial form texture of the South Railway Station in the form of red and black I-beam steel frames and partial glass roofs, and forms a landscape in the viewing square with steel frames and black sleepers engraved with the historical contents of the South Railway Station. The sculpture group forms a plurality of linear cantilever viewing platforms in the form of multiple railway crossings, paving with roadbed stone as the main filling material, and a dynamic wooden platform and a cantilever viewing platform to form a dynamic four-dimensional viewing space.

The eight-way wharf uses the existing steps and the structural parts of the remnants in the middle of the relics to form a viewing platform and a three-dimensional space with black I-beam steel frame, black sleepers and transparent glass, taking into account the various aspects of protection, safety, viewing and riverside landscap retaining the remains as much as possible while weakening the new design elements.

The train themed square and the liquefied gas terminal, in the form of a "fashion base", extend the railway relic culture of the train themed square with the adjacent liquefied ga terminal, and update the commercia connotation of the square to form the landscape system in this node. The sports facilities on the square, the train teahouse, the factory bar, the

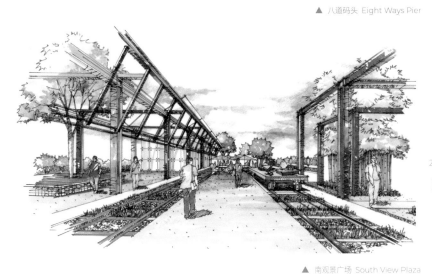

▲ 八道码头 Eight Ways Pier

▲ 南观景广场 South View Plaza

▲ 廊架 Gallery

nderwater performance stage and
e roadbed stone pavement and
her landscape elements enable the
ic landscape to evoke the sense
youthful vitality and the cultural
emory of the urban people.
e landscape structure stresses
ur points, one line, multiple
ers, highlighted treatment,
ear connection and spatial level.
construct the frame scene of the
site of the South Railway Station
d reproduce the rails and shipping
cks; Respect the current conditions

of the eight-way wharf, overhead
trestle and glasses, and the site is
treated with simplicity; Reconstruct
the train square and the liquefied gas
terminal, implant fashion activities
and extreme sports, and expand the
underground passage to connect the
square, river beach and dock to create
a fashion base in the region; Integrate
the bridgehead green land that returns
to nature with the southern suburb
garden; Link free linear water platform
and the riverside background forest
belt with the mountains and water

split by the newly-built avenue, the
multi-layered spatial landscape effect
comes from a rich, diverse and three-
dimensional landscape structure. The
fuzzy boundary, texture structure and
multi-level transition method make
up for the design defects left in the
road design, so that nature, road and
landscape form a harmonious system.

▲ 廊架 Gallery

▲ 观景平台 Viewing Platform

▲ 廊架 Gallery

▲ 休闲广场 Leisure Square

▲ 火车广场 Train Square

园林设计研究讲究"三脉"：文脉、水脉、绿脉。以体现三国故里文化内涵为突出特征的文脉，以立体、复合、多彩的植物搭配为突出特征的绿脉，以创造"亲民亲水"的休闲场所为主要水脉特征。三者互相叠加、渗透、交融，形成有特色、有内涵的和谐景观。

逍遥津文化公园是以满足城市居民日常游憩需求为主要功能，以绿色生态为主要形式的城市游憩公园，同时又以其独特的城市文化折射出城市形象和公园形象，吸引着周边地区的人们前来观光、游览、娱乐。

"Three veins" are stressed in landscape design study: cultural context, water vein and green vein. The cultural context characterized by the cultural connotation of the Hometown of the Three Kingdoms, the green vein characterized by three-dimensional, complex and colorful plants, as well as the water vein characterized by creating a "people-friendly and water friendly" leisure place, the three veins overlap, permeate and blend with each other, forming a harmonious landscape.

Leisure Ferry Cultural Park is an urban recreation park with the main function of meeting the daily recreational needs of urban residents, and takes green ecology as its main form. At the same time, its unique urban culture reflects the image of the city and the park, thus attracting people in the surrounding areas for sightseeing and entertainment.

合肥逍遥津文化公园
Hefei Leisure Ferry Cultural Park

设计时间：2008 年
项目委托：合肥市政府
项目地址：安徽 合肥
项目规模：58000 ㎡

Design time: 2008
Entrusting party: Municipal Government of Hefei
Project site: Hefei, Anhui
Project scale: 58,000 ㎡

▲ 亲水平台
Hydrophilic Platform

◄ 滨湖空间
Lakeside Space

◀ 亲水平台夜景
Hydrophilic Platform Night View

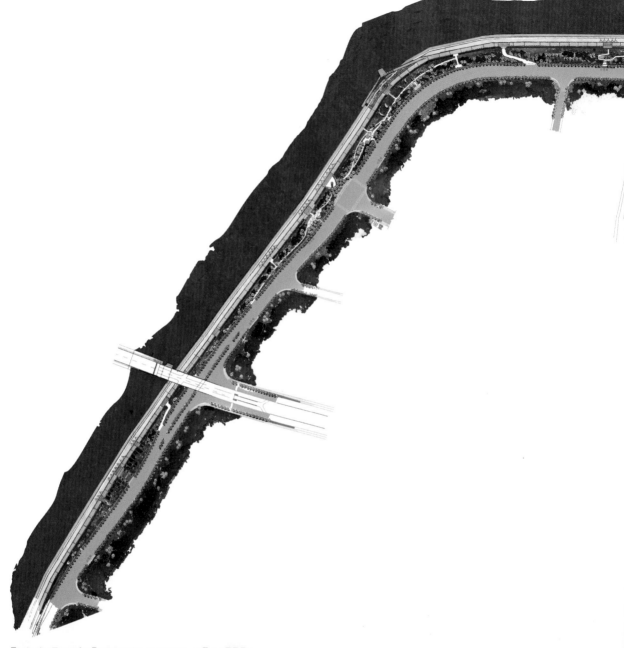

长沙浏阳河风光带
Changsha Liuyang River Scenic Area

设计时间：2007 年
项目委托：长沙市南湖新城工程建设指挥部
项目地址：湖南 长沙
项目规模：165000 ㎡

Design time: 2007
Entrusting party: Nanhuxincheng Engineering Construction Headquarters of Changsha City
Project site: Changsha, Hunan
Project scale: 165, 000 ㎡

日河风光带是体现长沙城市形象的重要景
走廊，是芙蓉区展现人文和环境品位、改
人居环境的重要滨水开放空间，是城区向
村延伸、乡村向城区渗透的一道风景线。

次设计赋予浏阳河风光带的主题是：浏阳
——"红色"景观长廊，以场所历史特征
基础、以富有时代感的形式予以表现，创
地区特色的唯一性。基于此，设计强化大
心目中浏阳河的"红色"印象，将其打造
中国第一条"红色"景观长廊。

本布局上从城市尺度分析浏阳河，我们可
确定重要的景观节点及标志点的位置，即
往入湘江的入江口及河道转弯的拐点处。

缩小到芙蓉区的尺度，根据对周边情况的分
析，并结合基地特征，确定每段的分主题。
从康乐路口至晚报大道路口为北段，表现如
火如荼的经济建设。从晚报大道路口至支路
二十九路口为中段，结合古汉路表达神秘的
湘楚文化。从支路二十九路口至咿呀台为南
段，结合"迎解"古樟表现革命历史。

Liuyang River Scenic Area is an important landscape corridor that reflects Changsha city image; it is an important waterfront open space that shows humanist and environmental taste of Furong District, and improves human settlement environment; it is a landscape line between the urban area and the countryside.

The theme of Liuyang River Scenic Area in the design is: Liuyang River: "Red" landscape corridor based on the historical characteristics of the place, expressed in the form of rich sense of the times, so as to create the uniqueness of regional characteristics. And based on this, the design strengthens the "red" impression of Liuyang River in the eyes of the masses and makes it the first "red" landscape corridor in China.

According to the analysis of the overall layout of Liuyang River in terms of urban scale, we can determine the location of the important landscape nodes and landmarks, which is, at the entrance of Xiangjiang River and the turning point of the river course. Reducing to the scale of Furong District and according to the analysis of the surrounding conditions and base characteristics, the sub-theme of each segment can be determined. The northern section is from the intersection of Kangle Road to that of Evening News Avenue, which showcases the prosperous economic construction. The middle section is from the intersection of Evening News Avenue to that of bypass No. 29 Road, which combines Guhan Road to embody Xiang-Chu culture. The southern section is from bypass No. 29 Road to Yiyatai, which combines Yingjie ancient camphor trees to demonstrate the history of revolution.

▶ 观景木平台
Viewing Wooden Platform

景观廊架
Landscape Gallery

滨河休闲带
Binhe Leisure Belt

主题广场
Theme Square

赛瑞景观二十周年作品特辑

设计方案将原有的裕湘纱厂和船舶厂区域进行统一规划。长沙的 LOSCAR 滨江城市景观区，是长沙城市商业空间网络结构中独一无二的商业空间节点，是一个以城市普通市民为主要服务对象的现代生活模式社区。在总平面中看似标准化实体空间形式像细胞一样的开放空间共生的形式，象征着工业化与信息化时代的交叉与融合。

The design proposal makes a combined planning of the former Yuxiang Cotton Mill and Shipyard. As the unique commercial space joint of its urban commercial space network, LOSCAR riverside urban scenic area in Changsha is a community of modern lifestyle serving common citizens in the city. In the general layout, it looks like a standard physical space, however, in fact, it is a cell-like open space of commensalism, which symbolizes the intersection and integration of industrialization and informatization.

长沙渔人码头
Fisherman Dock, Changsha

设计时间：2009 年
项目委托：长沙市政府
项目地址：湖南 长沙
项目规模：79000 ㎡

Design time: 2009
Entrusting party: Municipal Government of Changsha
Project site: Changsha, Hunan
Project scale: 79, 000 ㎡

▲ 总体 Overall

景观空间系统分为建筑空间、开放空间、
五大道和星光塔四个主要部分。其中建筑
间中包括实体建筑和围合空间，而开放空
中包括室外休闲、演示、活动、观景空间
素和植物系统。

The landscape space consists of
architectural space, open space, the 5
Avenue and Starlight Tower, of which
the architectural space is composed c
physical structure and enclosed space
the open space includes outdoor
entertainment, demonstration, activit
viewing space elements and plant
system.

◀ 船坞空间 Shipyard Space

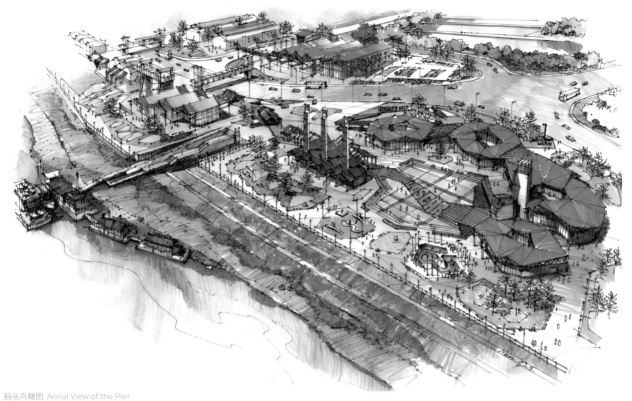

码头鸟瞰图 Aerial View of the Pier

第五大道 Fifth Avenue

山东潍坊白浪河

Bai Lang River, Wei Fang City, Shandong

设计时间：2005 年
项目委托：山东潍坊市政府
项目地址：山东 潍坊
项目规模：1000000 ㎡

Design time: 2005
Entrusting party: Municipal Government of Weifang City, Shandong Province
Project site: Weifang, Shandong
Project scale: 1, 000, 000 ㎡

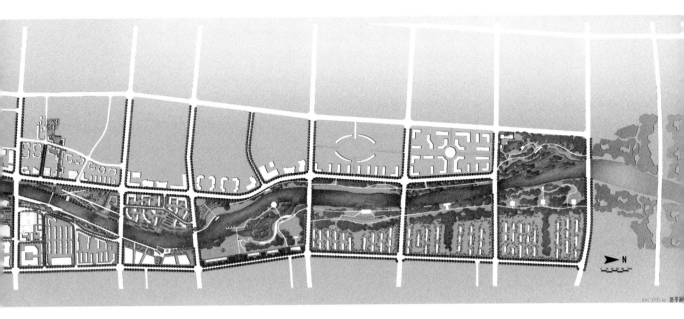

次规划设计将白浪河定位为可以与法国塞河、英国泰晤士河相媲美的国际一流滨水景线。以水为纽带贯穿潍坊市区南北，融合野与城市，包含生态保育、文化旅游、商贸易、观光休闲的主要功能，本案在国际标中荣获第一名。

方案设计中，沿白浪河构建了三个不同的域主题：自然生态（野生水体）、城市底（城河相际）、文化背景（田园水系）。方面将野生自然与富饶土地相联系，另一面在白浪河中心流域也就是河流与城市相之处，形成一个文化内涵丰富的滨河区域。河设计具有各种特色的区域和节点，力图清晰的规划和理念贯通于设计中，给游人来步移景异的景观体验，如从南面的水坝湿地、植物园，再到新旧 CBD，最后抵达部的运动和民俗园区。

The planning design positions Bai Lang River as the international first-class waterfront landscape comparable to Seine River in France and the Thames in U.K. It connects the north and the south of Weifang urban area linking with water, blends the countryside and the city, includes the main functions of ecological conservation, cultural tourism, commercial trade, sightseeing and leisure, and the design case won the first place in the international bidding.

In the design scheme, three different regional themes are established along Bai Lang River: natural ecology (wild water body), urban foundation (junction of city and river), as well as cultural background (rural water system). On the one hand, it connects wild nature

and rich land. On the other hand, it creates a riverside region with rich cultural connotations at central Bai Lang River where the river intersects with the city. And there are various regions and nodes with different characteristics along the river. It tries to make clear plan and concept in the design, brings the tourists a landscape experience of different views as one moving his/her steps, for instance, from the dam in the south to the wetland, botanical garden, to the old and new CBD, and ultimately to the sports and folk custom park in the north.

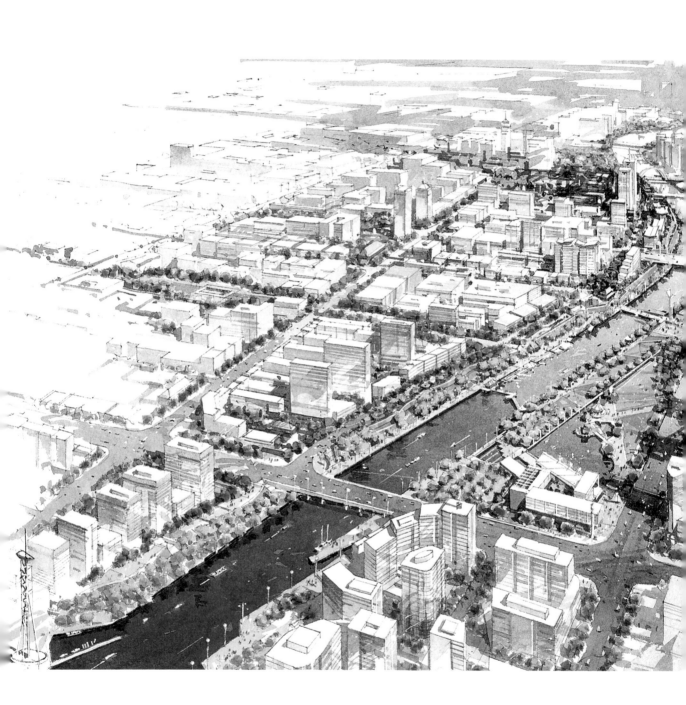

▲ 奎文门广场 Kuiwenmen Square

▲ 滨河步道 Riverfront Trail

▲ 教育主题公园 Educational Theme Park

▲ 水晶街区 Crystal District

▲ 鸟瞰图 Aerial View

▲ 鸟瞰 Aerial View

2000-2020 赛瑞景观项目年鉴

A
安徽柏景湾花园
安徽宿州万达广场
安乡山湖海上城园林
澳门柯维纳

B
蚌埠恒大御景湾
包头恒大名都
包头万达广场
北辰三角洲项目 D2 区
北京耕天下
北京首创成都胜利村项目
本溪恒大绿洲
碧桂园潼湖创新小镇首开区
滨河海马公园项目展示区及园区
滨江天地花海项目
博雅滨江四期
博雅公馆
博雅花园

C
长安国际广场
长春高新海容广场
长春恒大御景
长房·潭房时代公馆
长房宁乡项目
长房宁州府一期
长房西府项目
长房雍景湾
长江湘江大道
长沙八方小区
长沙北辰三角洲
长沙格兰小镇
长沙国际会议展览中心
长沙恒大雅苑
长沙环球世纪未来城
长沙金色华庭环境扩初
长沙浏阳河风光带
长沙市芙蓉区文化活动中心
长沙市浏阳河文化公园
长沙一馆三中心
长沙渔人码头
常德恒大华府
常州阳湖名城
郴州公园九里
郴州金科城
成都奥特莱斯购物公园
成都翡翠华庭
成都恒大御景半岛
成都华侨城纯水岸
成都佳兆业八号
成都金科牧马山
成都蓝光云鼎
成都领地国际广场
成都领地中心
成都润扬泥巴沱
成都威斯顿联邦大厦
城市山谷巴黎道景观改造
赤湾项目地铁上盖
重庆地王广场环境
重庆奉节·山湖海上城一期
重庆恒大帝景
重庆恒大金碧天下
重庆环球广场
重庆金科时代中心
重庆金科王府公园
重庆金科西永 L48-5 地块
重庆金科阳光小镇
重庆开县财富中心
重庆开县东部酒店
重庆万州金科
重庆渝海城北区高层屋顶花园
重庆长寿金科世界城
春华健康城一期

D
大厂田各庄 78 亩
大理海东下和国际艺术小镇
大理万德名邦 55 下和国际艺术小镇
大荔同州北路市政绿化带
大连软件园
大庆北国之春梦幻城
大庆恒大绿洲
大同恒大绿洲
大同阳高温泉旅游度假区
儋州恒大海花岛
儋州恒大金碧天下
得天和苑项目
东莞翡翠山湖小高层
东莞凤岗锦绣华庭
东莞凤岗锦绣华庭
东莞凤岗幼儿园
东莞汇景豪庭二期
东莞金地·博登湖三期
东莞锦江丽园
东莞凯伦花园
东莞丽水佳园
东莞牛杨地块项目
东莞时代金融中心项目
东莞市凤岗镇玉泉路街心绿地
东莞松湖朗苑
东莞天安数码城
东莞星城玉珑湾
东莞星海国际花园
东莞盈丰商住中心园林
东莞御花园 B 住宅区
东江源三百山温泉度假公园
东营恒大棕榈岛

F
沣东自贸产业园三期
佛冈恒大金碧山庄
佛山金碧海岸样板房景观
佛山龙光·玖龙郡
佛山南海数码新城
佛山天安南海数码新城
富山御苑

G
赣州黄金时代
广东佛山南海天安科技数码城
广东万鑫五洲风情
广晟海韵君兰
广州白天鹅花园
广州北大资源博雅 1898
广州博雅花园
广州大道北项目
广州番禺动漫产业园
广州番禺锦绣香江
广州光大花园
广州恒大御景湾
广州宏润阳光花园
广州市番禺节能科技园
广州天安科技创新大厦
广州万达主题乐园景观梦幻花园
广州新白云国际机场
广州休闲小楼
贵港东湖公园
贵港国际生活港
贵阳恒大绿洲
贵阳太立常青藤花园
贵阳新太乙在水一方
贵州时光
桂林电子商城天悦城
桂林融创万达文化旅游城·栖霞府北苑

H
哈尔滨恒大名都
海南海花岛 1 号桥
海南南美假日
杭州宝力君汇中央
杭州华立东方俊园
杭州中兴和园山居
杭州中兴花园
浩创果岭上院项目园林
合肥都市森林
合肥逍遥津文化公园
河北建设福美 D 区
河北张家口尚峰国际
河南洛阳东方今典三地块
河南尚城英郡
河源万隆公园壹号 2 期
河源响水森林体育公园
衡阳蒸水河风光带
衡阳中泰天境
洪湾旧村改造项目 01、02 地块
鸿溪花园
呼和浩特恒大翡翠华庭
湖北京山国际温泉度假村
湖南芙蓉国豪廷大酒店
湖南华银天际
湖南三一街区三期
湖南省博物馆
湖南亚大时代
湖南宜章项目东地块
湖南岳泰新空间
华创·津南科技园
华创动漫园二期
华侨城·坪山龙田文旅小镇近期范围
华侨城创想大厦
华侨城金山湖壹号
华侨城香山美墅
淮北恒大名都期
淮北恒大雅苑
会展湾南岸广场（5-02 地块）
惠东二期 12#-17# 架空层
惠州东江熙园项目一期、二期
惠州汤臣花园
惠州星河丹堤

J
吉安恒大帝景
济南恒大绿洲
济南恒大裴翠华庭
济南万达城
建发连江青塘项目
建发宁德天行泱著项目
建发泱郡项目
建瓯建发·玺院
建瓯市建发玺院项目大盘区
建瓯市下水南片区
江门世纪商业街
江苏佳和水岸
江苏无锡金色水岸
江苏阳光蠡湖 1 号
江西鑫新滨江西路
江西中江紫金城
江阴蓝天豪庭
交投置地·承平盛世项目
焦作城际花园
揭阳恒大华府
金方圆 3 号地块嘉润公馆
金方圆广场
金科公园王府
金升地产九曲岭
金仕顿大酒店屋顶花园温泉
金源小区
晋州欧景城
景峰广场 2 期
景园·美麓城
九江庐山豪庭
俊怡九江庐山御府

K
开县金科天湖印项目
康伦佳境天成花园
康美健康城天麓苑（C2、C4 地块）
昆明华侨城高尔夫酒店
昆明华侨城圣托里尼酒店
昆明佳兆业城市广场
昆明万达滇池卫城·蓝岸

L
兰州恒大名都
廊坊恒大名都
连云港恒大城
廉江悦府
聊城恒大名都
陵水红磡香水湾金缔度假酒店
领地·凯旋国际公馆
领地中心
龙光玖誉花园
龙光顺德五沙珑滨熙园
罗浮山健康生态城

M
茂名名雅花园
眉山凯旋国际公馆
梅州石磊高尔夫球会
绵阳仙海长虹创新研发园

N
南昌万科洪坊
南充清泉滨江天地
南京句容恒大雅苑
南京聚宝山庄
南京万达华府二期
南京阳光聚宝山庄
南宁八桂绿城二期
南宁翡翠园
南宁天健世纪花园
南平东坑片区 1# 地块
内罗毕项目

波国际贸易展览中心景观改造
波恒大山水城
波亲亲家园
波盛世天城二期

博·金城珑园一期
顶山朗曼 0375·首府
山益田共和城邦项目 2 号地

齐哈尔恒大名都
齐哈尔名都
西中泰峰境
水湾山庄（三期）项目园林景观工程
皇岛恒大城
成达光明项目
岛北胶州湾新产业基地
岛集力豪庭
岛千千树项目
岛唐岛湾游艇会
岛远洋公馆
岛卓越大厦
岛卓越蔚蓝群岛
冈花苑二号院
远清晖路 50 亩项目
远威龙新城环境设计
州恒大御景湾
清外滩
清潇湘
州恒大都市广场

卫槟榔河
卫景园城
卫三美湾度假酒店
卫同济科技园
卫阳光海岸
卫卓达热带雨林高尚社区
禾花园
东大塘泰华城
东潍坊白浪河
湖廉江悦府
东长平公园
忆·爱琴湖主题公园
前海人寿医院项目西区
川宝安 76 区楼盘
川城市硅谷
川城市山谷花园三期
川大工业区行政商务区
川德业基德邻雅筑
东埔碧海蓝天明苑
川富通丽沙花都
川富通丽沙花都城市之光
川富通瑞翔居、瑞尚居
川观澜高尔夫假日赛维
川冠懋华苑花园
川国际和平城
川和兴花园二期
鹤围村文化体育广场
川鸿基新鸿一期
川鸿荣源环境投标
川浩源 A 区四期
川华侨城布兰谷地
川华侨城纯水岸
川华侨城栖湖花园
川华侨城香山西街
川皇城皇御苑 B 区
川金成时代家园
川葵涌海鲜街

深圳龙珠医院改扩建项目
深圳绿景红树华府
深圳南山豪庭
深圳全海花园
深圳沙嘴村景观改造工程
深圳厦门海沧工业园
深圳深茂威凤山庄
深圳深业新岸线
深圳神州通水澜山
深圳世界之窗外观改造
深圳市管办大楼
深圳泰富华天峦湖
深圳天安创新科技广场
深圳天安高尔夫天城
深圳天safe数码城
深圳万科四季花城
深圳文雅豪庭
深圳西乡迎大运街道分赛场
深圳仙泉山庄
深圳心湖路环境改造
深圳新城宾馆
深圳新鸿花园
深圳馨苑园林
深圳星河丹堤
深圳泽润翰园
深圳招商华侨城会展湾南岸广场
深圳中航格澜阳光
深圳中深三九泰苑
深圳中心城泽润瀚园
深圳紫郡
沈阳恒大江湾
沈阳恒大名都二期
沈阳泰宸
石家庄恒大名都
石家庄恒大雅苑
石家庄恒大御景半岛
石家庄欧景华庭
石家庄雅苑
世界城项目展示区
世林弘府
顺德华侨城天鹅湖
顺德佳兆业金域天下
松原市恒大御景湾

T
太阳新城
太原恒大绿洲
太原恒大小店区康宁街
太原茂业天地
太原南堰村
唐山凤凰新城地块
天地源·曲江香都 B 区
天健工业园片区更新单元
天健惠州项目
天津红桥文园
天津麦收
天津卫道春天
天津新中大春天
天津新中大月光花园
天津御景半岛

W
万州金科观澜
潍坊东方天韵
潍坊恒大翡翠华庭
乌兰浩特恒大绿洲
无锡海岸花园
无锡恒大城
无锡蠡湖新城一地块
无锡太湖金色水岸别墅

无锡鑫湖一号水域城邦
吴江金科廊桥水岸
武汉大华南湖公园世家
武汉恒大城二期
武汉华侨城
武汉中央商务区宗地

X
西安国际广场
西安恒大绿洲
西安汇通太古城
西安金地·西洋路中学
西安泾河小镇
西安人民大厦
西安樱花苑
西部云股二期
西宁恒大名都
西双版纳湿地公园
香蜜公园
湘潭长房·万楼公馆
襄阳隆中古镇
新疆凯旋公馆和眉山凯旋公馆
新疆领地凯旋国际公馆
新疆三泰左岸明珠花园小区
新乡恒大雅苑
信阳正商红河谷
邢台恒大城 A 地块
邢台恒大帝景
许昌恒大绿洲

Y
雅安领地凯旋帝景
烟台银和怡海天越湾工程
延平建发玺院一区
阳江华越御景豪庭
杨凌恒大城
伊宁恒大绿洲
益田益科大厦
雍景·上河城
永安建发·玺院
永安建发大区
御园·桃李春风
远洋盛平项目一期
岳阳恒大名都
岳阳金桥花园第三期
岳阳天灯咀项目公园
悦珑湾花园
越南胡志明市公园景观改造
云南官房云秀康园
云南华侨城高尔夫会所
云南银德香谢丽园
运城恒大绿洲

Z
湛江观海长廊环境设计方案
招商局光明科技园三期 A4 项目
肇庆龙光·玖龙山
浙江嘉兴亚厦风和范
镇江万科蓝山
郑州海马公园一、二、三期
郑州海马青风公园
郑州恒大名都
郑州华侨城中原项目
郑州建正东方中心
郑州静泊山庄
郑州开元城
郑州市静泊山庄
中房瑞致小区
中广天择总部基地
中国深圳金地集团大厦 101 栋改建

中国注册会计师北京培训基地
中建·益阳城市学院地块
中科亿方智汇产业园
中山翠景项目
中山富和城东花园
中山富和长江北
中山富鑫花园
中山海雅
中山名都花园
中山润达华府一区
中泰天成臻府住宅
中央城·中央大街
中冶口岸大厦项目
周庄新市镇·展示区
珠海斗门镇龙山湖
珠海金海景花园
珠海蓝湾半岛
珠海亮华南村豪苑
珠海中冶口岸大厦
自贡贡山壹号
自贡雍锦湖 A 区

* 中文字符按拼音首字母顺序排列

2000-2020 CSC PROJECT YEARBOOK

A

Administrative and Business District of Shenzhen
Industrial Distric
Ageas Lisha Flower City in Shenzhen
Ageas Lisha Flower City Light of the City in
Shenzhen
Ageas Ruixiang Mansion and Ruishang Mansion in
Shenzhen
Anhui Bojingwan Garden
Animation Industrial Park in Panyu, GUangzhou
Avenue North Project in Guangzhou
AVIC Gran Sunshine in Shenzhen

B

Bagui Green City Phase II in Nanning
Bailang River in Weifang, Shandong
Baoli Junhui Committee in Hangzhou
Beida Resources Boya 1898 in Guangzhou
Beijing Initiation Chengdu Victory Village Project
Blue Bay Peninsula in Zhuhai
Blue Light Clout Tripod Community in Chengdu
Blue Sky Mansion in Jiangyin
Boden Lake Phase III in Dongguan Jindi
Boya Garden
Boya Garden in Guangzhou
Boya Mansion
Boya Riverside Phase IV
Bran Valley in Overseas Chinese Town, Shenzhen
Brilliant Blue Islands in Qingdao
Bule Ocean and Blue Sky Bright Garden in
Dongpu, Shenzhen

C

Central City Central Street
Central Plains in Overseas Chinese Town,
Zhengzhou
Chanfine Ningxiang Project
Chanfine Ningzhou Mansion Phase I
Chanfine Tanfang Time Mansion
Chanfine West Mansion Project
Chanfine Yongjing Bay
Chang'an International Plaza
Changchun High-Tech Hairong Plaza
Changping Park in SHantou
Changsha Fisherman's Wharf
Changsha Furong District Cultural Activity Center
Changsha Global Century Future City
Changsha Golden Courtyard Environment
Expansion
Changsha International Convention and Exhibition
Center
Changsha Liuyanghe River Cultural Park
Chengdu Evergrande Yujing Peninsula
Chengdu Outlets Shopping Park
Chengdu Overseas Chinese Town Pure Coast
China Certified Public Accountant Beijing Training
Base
China Merchants Group Bright Science and
Technology Park Phase III A4 Project
Chongqing Diwang Square Environment
Chongqing Yuhai Chengbei District High Rise Roof
Garden
Chunhua Healthy City Phase I
City Above Mountain, Lake and Ocean Phase I in
Fengjie, Chongqing
City Gardens above Mountain, Lake and Ocean in

Anxiang
Coastal Garden in Wuxi
Commercial and Residential Central Garden in
Dongguan Yingfeng
Corvina, Macau
Courtyard No.2 of Flower Court
CSCEC Yiyang City College Plot
Cuijing Project in Zhongshan
Cultural Sport Square in Hewei Village, Shenzhen

D

Dali Haidong Xiahe International Art Town
Dali Tongzhou North Road Municipal Green Belt
Dali Wonder Eminent City 55 Xiahe International
Art Town
Dalian Software Park
Dealskey Delin Elegant Building in Shenzhen
Delta Project in Changsha
Detian Heyuan Project
District D2 of Delta Project in Beichen
Dongfang Jindian Lot 3 in Luoyang, Henan
Dongjiang Xiyuan Project Phase I and Phase II in
Huizhou
Dream City of Spring in the North Country in
Daqing

E

East Plot of Yizhang Project in Hunan
Elegant Court in Shijiazhuang
Emerald Court in Chengdu
Emerald Mountain Lake Small High-Rise in
Dongguan
Eminent Metropolis Garden in Zhongshan
Eminent Metropolis in Qiqihar
Energy Saving Science and Technology Park in
Panyu, Guangzhou
Environment Design of Weilong New Town in
Qingyuan
Environmental Design Plan of Sea Viewing Corridor
in Zhanjiang
Environmental Reconstruction of Xinhu Road in
Shenzhen
European Scene City in Jinzhou
Everbright Garden in Guangzhou
Evergrande City in Lianyungang
Evergrande City in Qinhuangdao
Evergrande City in Wuxi
Evergrande City in Yangling
Evergrande City Plaza in Quanzhou
Evergrande City Plot A in Xingtai
Evergrande Elegant Court in Changsha
Evergrande Elegant Court in Huaibei
Evergrande Elegant Court in Jurong Nanjing
Evergrande Elegant Court in Shijiazhuang
Evergrande Elegant Court in Xinxiang
Evergrande Emerald Court in Huhhot
Evergrande Emerald Court in Jinan
Evergrande Emerald Court in Weifang
Evergrande Eminent Metropolis in Baotou
Evergrande Eminent Metropolis in Harbin
Evergrande Eminent Metropolis in Huaibei
Evergrande Eminent Metropolis in Langfang
Evergrande Eminent Metropolis in Lanzhou
Evergrande Eminent Metropolis in Liaocheng
Evergrande Eminent Metropolis in Qiqihar
Evergrande Eminent Metropolis in Shijiazhuang

Evergrande Eminent Metropolis in Xining
Evergrande Eminent Metropolis in Yueyang
Evergrande Eminent Metropolis in Zhengzhou
Evergrande Eminent Metropolis Phase II in
Shenyang
Evergrande Flower in Ocean Island in Danzhou
Evergrande Golden World in Chongqing
Evergrande Inxuriant Mansion in Changde
Evergrande Inxuriant Mansion in Changde
Evergrande Jinbi Mountain Villa in Fugang
Evergrande Landscape City in Ningbo
Evergrande Oasis in Benxi
Evergrande Oasis in Daqing
Evergrande Oasis in Datong
Evergrande Oasis in Guiyang
Evergrande Oasis in Jinan
Evergrande Oasis in Taiyuan
Evergrande Oasis in Ulanhot
Evergrande Oasis in XI'an
Evergrande Oasis in Xuchang
Evergrande Oasis in Yining
Evergrande Oasis in Yuncheng
Evergrande Palm Island in Dongying
Evergrande Phase II in Wuhan
Evergrande River Bay in Shenyang
Evergrande Royal Scene in Chongqing
Evergrande Royal Scene in Ji'an
Evergrande Royal Scene in Xingtai
Evergrande Splendor in Danzhou
Evergrande Yujing in Changchun
Evergrande Yujingwan in Bengbu
Evergrande Yujingwan in Guangzhou
Evergrande Yujingwan in Quzhou
Evergrande Yujingwan in Songyuan
Excellence Building in Qingdao
Exhibition Area and Park of Riverside Seahorse
Park Project
Expo Bay South Coast Plaza (Lot 5-02)
Expo Bay South Coast Plaza in Merchant Oversea
Chinese Town, Shenzhen

F

First Development District of Tonghu Innovation
Town of Country Garden
Free Trade Industrial Park Phase III in Fengdong
Fuhe East City Garden in Zhongshan
Fuhe North of Changjiang River in Zhongshan
Fuxin Garden in Zhongshan

G

Garden of Haochuang Green Upper Courtyard
Project
Gengtianxia Community in Beijing
Global Plaza in Chongqing
Golden Age of Ganzhou
Golden Bridge Garden Phase III in Yueyang
Golden Coast in Wuxi, Jiangsu
Golden Coast Villa in Taihu Lake, Wuxi
Golden Fangyuan Square
Golden Sea View Garden in Zhuhai
Golf Culb in Overseas Chinese Town, Yunnan
Gongshan No.1 in Zigong
Gran Town in Changsha
Green Scene Red Trees Inxuriant Mansion in
Shenzhen
Green Space at Center of Yuquan Road in

Shenzhen Guanmao Inxuriant Garden
Shenzhen Hongji Xinhong Phase I
Shenzhen Hongrongyuan Environmental Bidding
Shenzhen Huahaoyuan District A Phase IV
Shenzhen Nanshan Mansion
Shenzhen Royal City Royal Mansion District B
Shenzhen Shazui Village Landscape
Reconstruction Project
Shenzhen Shenmao Weifeng Villa
Shenzhen Shenye New Coastline
Shenzhen Shenzhou Tongshui Lanshan
Shenzhen Taifu Huatianluan Mountain
Shenzhen Urban Management Office Building
Shenzhen Urban Silicon Valley
Shenzhen Urban Valley Garden Phase III
Shijiazhuang Evergrande Yujing Peninsula
Shilinhongfu Mansion
Songhu Langyuan in Dongguan
South America Holiday in Hainan
South Area of Xiashui, Jianou
Splendid Court in Dongguan Fenggang
Splendid Court in Dongguan Fenggang
Splendid Xiangjiang in Panyu, Guangzhou
Spring of Weidaoguo in Tianjin
Subway Upper Lid of Chiwan Project
Sun New Town
Sunshine Coast in Sanya
Sunshine Treasure Valli in Nanjing
Swan Lake in Overseas Chinese Town, Shunde

T
Taichen in Shenyang
Taihua City in Dayong, Shandong
Taili Ivy Garden in Guiyang
Territory Center in Chengdu
Territory Centre
Territory International Plaza in Chengdu
Territory Triumph International Mansion
Thousands of Trees Project in Qingdao
Tian'an Digital City in Dongguan
Tian'an Digital City in Shenzhen
Tian'an Golf Park in Shenzhen
Tian'an Innovation Technology Plaza in Shenzhen
Tian'an Nanhai Digital New Town in Foshan
Tian'an Nanhai Technology Digital City in Foshan,
Guangdong
Tian'an Science and Technology Innovation
Building in Guangzhou
Tiandengzui Project Park in Yueyang
Tiandi Source Xiangjun District B in Qujiang
Tianjian Century Garden in Nanning
Tianjian Huizhou Project
Tianjin New Zhongda Moon Light Garden
Tianjin New Zhongda Spring
Tianjin Yujing Peninsula
Time Financial Centre Project in Dongguan
Tomson Garden in Huizhou
Tongji Science Park in Sanya
Treasure Villa in Nanjing
Triumph International Mansion in Meishan

U
Urban Forest in Hefei

V
Vanke Blue Mountain in Zhenjiang
Vanke Four Seasons Flower City in Shenzhen

W
Wanda City in Jinan

Wanda Dian Lake Acropolis in Kunming Blue Bank
Wanda Plaza in Baotou
Wanda Plaza in Suzhou, Anhui
Wanxin Wuzhou Amorous Feelings in Guangdong
Wanzhou Jinke in Chongqing
Wanzhou Jinke Mission Hills
Wenya Elegant Mansion in Shenzhen
West Area of Shaoguan Qianhai Life Insurance
Hospital Project
Western Cloud Stock Phase II
Weston Federal Building in Chengdu
Wheat Harvest in Tianjin
White Swan Garden in Guangzhou
World City Project Exhibition Area
Wuhan CBD Parcel
Wuhan Dahua Nanhu Park Family
Wuxi Xinhu No.1 Water City State

X
Xi'an Cherry Garden
Xi'an Jindi Xifeng Road Middle School
Xi'an Jinghe Town
Xi'an People's Building
Xiamen Haicang Industrial Park in Shenzhen
Xiangjiang Brood Road of Changjiang River
Xiangmi Park
Xiangshan Villa in Overseas Chinese Town
Xiangshan West Street in Overseas Chinese Town,
Shenzhen
Xiangyang Longzhong Ancient Town
Xianquan Villa in Shenzhen
Xiaoyaojin Cultural Park in Hefei
Xinghe Dandi in Huizhou
Xinghe Dandi in Shenzhen
Xinhong Garden in Shenzhen
Xinjiang Territory Triumph International Mansion
Xinjiang Triumph Mansion and Meishan Triumph
Mansion
Xinyang Zhengshang Red River Valley
Xinyuan Garden in Shenzhen
Xishuangbanna Wetland Park
Xixiang Yingdayun Street Sub Venue in Shenzhen

Y
Ya'an Territory Triumph Royal Scene
Yacht Club of Tangdao Bay in Qingdao
Yada Time in Hunan
Yanggao Hot Spring Resor in Datong
Yanghu Eminent Metropolis in Changzhou
Yangjiang Huayue Imperial Court
Yantai Yinhe Yihai Tianyue Bay Project
Yaxia Fengye Court in Jiaxing, Zhejiang
Yike Building in Yitian
Yong'an Jianfa Royal Land
Yonan Jianfa District
Yongjin Lake District A in Zigong
Yongjing Shanghe City
Yuannan Yinde Xiangxie Liyuan
Yuanyang Shengping Project Phase I
Yuelong Bay Park
Yuetai New Space in Hunan
Yulong Bay in Dongguan Xingcheng
Yuyuan Taoli Spring Wind

Z
Zerun Hanyuan in Shenzhen
Zhengzhou Seahorse Green Wind Park
Zhengzhou Seahorse Park Phase I, II, III
Zhongfang Ruizhi Community
Zhongguang Tianze Headquarters Base

Zhongjiang Purple Golden City in Jiangxi
Zhongke Intelligent Aggregation from All
Directions Industrial Park
Zhongshan Haiya
Zhongshen Sanjiu Taiyuan in Shenzhen
Zhongtai Heaven in Hengyang
Zhongtai Mountain Scene in Qianxi
Zhongtai Tiancheng Zhenfu Mansion
Zhongxing Garden in Hangzhou
Zhongxing Heyuan Mountain Residence in
Hangzhou
Zhouzhuang New Town Exhibition Area

* 英文字符按项目名称首字母顺序排列
Arranged in alphabetical order

后记

文：贾罡
赛瑞景观执行董事

赛瑞景观成立于 2000 年，是伴随国家改革开放、大力推进城市化、大力发展房地产的大时代而发展起来的最早一批景观设计公司，至今整整二十年。赛瑞的发展大约经历了三个阶段：2000-2004 年的初创期，2005-2015 年的发展壮大期，2016 至现在的蓬勃发展、积极进取期，公司由初创期的几个人发展到今天 200 多人，并成为国内颇具影响力的景观设计公司，取得了令人骄傲的成绩。

赛瑞景观经过二十年的发展，形成了自己"兼容并蓄、鼎新独到、匠心规制、至臻服务"的独特的企业文化。"博观而约取，厚积而薄发"，是赛瑞人的精神特质，二十年磨一剑，我们结集出版《新共享景观》一书，本书既是赛瑞这二十年的成果回顾与总结，也是赛瑞人在互联网与共享经济等新经济模式下，进行城市重构、解决环境与人的关系的经验分享。我们提出"新共享景观"设计理念，为未来景观设计探索出一个新的思考方向。新的理念仍需要我们在工作中努力实践，我们期待为中国城市建设奉献更多适应时代发展的作品。本书也是赛瑞一次全方位的检阅、剖析和反思，希冀推动未来十年的持续成长。

过去的二十年，赛瑞人始终坚持以建设美好人居环境为使命，风雨同舟、勠力同心、潜心笃志、开拓进取，感恩你们 —— 一路同行的伙伴们！

过去的二十年，合作伙伴的大爱、信任与支持，助力我们不断突破自我、迎接挑战、取得成绩，感恩你们，我们的甲方朋友们！

过去的二十年，得力于业内外同仁们友情相随、鼎力相助、协作共赢，感恩你们，亲爱的朋友们！

同时，在这本《新共享景观》的编撰过程中，得到了许多合作企业和朋友们的支持，感谢诸位朋友的鼎力支持！

在本书与读者见面时，我们特别感谢孟建民先生、陈跃中先生在百忙之中拨冗作序，勉励后学。

感谢对本书策划、出版工作提供诸多帮助的中国建筑出版传媒有限公司！

感谢对本书提出修改意见，提供过帮助和支持的所有专家、学者和设计师！

感谢为本书提供大量精美实景图片的摄影师和为本书辛勤工作的编辑团队！

时光清浅，岁月嫣然，携一缕感悟于流年，那些镌刻在生命韵律中的温暖和感动，是生活留给我们的幸福痕迹，祝赛瑞人二十生辰快乐！

贾罡

2020 年 2 月 17 日于墨尔本

POSTSCRIPT

Writer: Gang Jia
Executive Director of CSC

Founded in 2000, CSC Landscape strives to develop among the earliest batch of landscape design companies benefited from the reform and opening up, vigorous promotion of urbanization and rapid development of real estate. CSC has gone through about three stages of development: the start-up period from 2000-2004, the development and growth period from 2005-2015, and the vigorous and aggressive development period from 2016-now. The company, with several persons initially to over 200 recruited today, has grown to be a top-ranking and influential landscape design company in China, and achieved great accomplishments.

CSC, after two decades of development, has formed its own unique corporate cultures of "inclusiveness, innovation, inventiveness, and perfectionism".

Characterized with "long-time integration, well-prepared accumulation," CSC personnel, with two decades' continued efforts, have collected multiple works and published the *New Shared Landscape*, a book reviewing and summarizing the results gained by CSC over the course of its development, and sharing the experience of CSC personnel in urban reconstruction and solutions to the relationship between environment and people under the new economic model such as Internet and sharing economy. We put forward the design concept of *"New Shared Landscape"* to explore a new thought for future landscape design. In the days ahead, we will step up our efforts to put the new concept into practice in our work, and we look forward to introducing more works that adapt to the development of the times for the construction of Chinese cities. This book, a comprehensive review, analysis and reflection of CSC, aims to promote its continued growth in the next decade.

Over the past two decades, CSC personnel have always adhered to the mission of "building a good living environment", and followed the principles of "unity, dedication, and pioneering". We sincerely thank all of our friends along the way.

Over the past two decades, it is the great love, trust and support of our partners that help us make constant breakthroughs, meet challenges and score achievements. Thank you so much, our Party A friends!

Over the past two decades, we have benefited from the friendship and cooperation of counterparts in and out the industry. Thank you so much, our intimate friends!

Simultaneously, we have received the support of many cooperative companies and friends in the compilation process of this *New Shared Landscape*. Thank you for your full support!

In addition, we especially thank Mr. Meng Jianmin, and Mr. Chen Yuezhong for their joint preface for encouraging younger leaders despite busy schedules. Thank you so much!

Thanks to China Architecture Publishing & Media Co., Ltd. who has provided great help in the planning and publishing of this book!

Thanks to all experts, scholars, and designers who have provided suggestions, help and support for this book!

Thanks to the photographers who have shot a large number of beautiful real-world scene pictures in this book and the hard-working editorial team!

Time flies, but we remain beautiful memories. With perception over time, the warmth and moving in the life are what make us feel happy. Wish CSC personnel a happy twentieth anniversary!

Jia Gang

February 17, 2020 in Melbourne

深圳总部办公区
Shenzhen headquarters office area

加拿大赛瑞（CSC）景观设计顾问公司

深圳市赛瑞景观工程设计有限公司

深圳总部：深圳市南山区侨香路 4060 号花样年香年广 B 座 301 - 304

0755 - 83306526

传真：0755 - 86096632

深圳公司邮箱：cscinfo@vip.163.com

西安赛瑞景观工程设计有限公司

西安分公司：西安高新区丈八一路汇鑫 1BC707

029-81024882

传真：029-81024877

西安公司邮箱：sairuicsc@163.com

Canada CSC Landscape Design Consultants

Shenzhen CSC Landscape Design Engineering Co., Ltd.

Shenzhen Address: 301 - 304, Block B, Future Plaza, 4060 Qiaoxiang Road, Nanshan District, Shenzhen

0755 - 83306526

FAX: 0755 - 86096632

Shenzhen Company Email: cscinfo@vip.163.com

Xi'an CSC Landscape Engineering Design Co.,Ltd

Xi'an Address: Xi 'an high - tech zone zhangs increases hui xin IBC C building, room 707

029 - 81024882

FAX: 029 - 81024877

Xi'an Company Email: sairuicsc@163.com

www.csclandscape.com

图书在版编目（CIP）数据

新共享景观 / 赛瑞景观编著 . － 北京： 中国建筑工业出版
，2020.7
ISBN 978-7-112-25258-9

Ⅰ.①新… Ⅱ.①赛… Ⅲ.①城市景观－景观设计－ 研究
.① TU984.1

中国版本图书馆 CIP 数据核字 (2020) 第 109088 号

誉主编：廖文瑾 贾 罡
行主编：贾 罡 丁 炯
起策划：深圳市赛瑞景观工程设计有限公司
辑：欧阳霞
术编辑：李 瑜
目摄影：杨 磊（琢墨摄影） 方 健（思康建筑摄影） 林 绿 聂 凡
文翻译：多才多译（北京）翻译有限公司

任编辑：唐 旭 吴 绫
字编辑：李东禧 孙 硕
任校对：张惠雯

新共享景观
赛瑞景观 编著
*
中国建筑工业出版社出版、发行（北京海淀三里河路9号）
各地新华书店、建筑书店经销
恒美印务（广州）有限公司印刷
*
开本：850×1168毫米 1/16 印张：20¼ 字数：400 千字
2020 年 7 月第一版 2020 年 7 月第一次印刷
定价：258.00 元
ISBN 978-7-112-25258-9
（36032）